AMERICAN CAVES and CAVING

AMERICAN CAVES and CAVING

*Techniques, pleasures, and safeguards
of modern cave exploration*

William R. Halliday, M.D.

*Former Director, Western Speleological Survey
President, Western Speleological Foundation*

REVISED EDITION

BARNES & NOBLE BOOKS
A DIVISION OF HARPER & ROW, PUBLISHERS
New York, Cambridge, Philadelphia
San Francisco, London, Mexico City
São Paulo, Sydney

First BARNES & NOBLE BOOKS edition published 1982.

ISBN 0–06–463556–2 (previously ISBN 0–06–011747–8)

82 83 84 85 86 10 9 8 7 6 5 4 3 2 1

Preface to the Revised Edition

Since the first edition, new technology and increased knowledge have greatly advanced speleology throughout North America. New ropes, new lights, new climbing gear, new knowledge—all these make today's caving a marvel even to those of us who have watched it grow. Entire new groups of vertical cavers have appeared, organized, and begun to publish newsletters within the National Speleological Society. The result has been success beyond our fondest hopes. The map of Mammoth Cave now extends far beyond Mammoth Cave Ridge and Flint Ridge. Lesser caves disdained a generation ago now are among the world's longest.

But the basic principles I outlined in 1974 are remarkably unchanged. The revisions in this edition concern details—details that make today's caving safer, more flexible, and even more delightful than that of the 1970s.

To each reader, good caving!

W. R. H.

Seattle
November 1981

Contents

Introduction

North American caving has burst national boundaries. Modern highways and twentieth-century technology have multiplied many times man's ability to respond to the lure of caves, and three decades of systematic studies have skyrocketed the known lengths of Stygian passageways. Throughout our continent, countless thousands increasingly seek the uniqueness of these passageways for the pleasure of exploration and reexploration, for scientific studies, or for the renewal of city-jaded souls.

But, with a few dramatic exceptions, many of even our most experienced cavers fail to appreciate the magnificent variety of our caves. Many traveling spelunkers expect those of distant regions to be mere facsimiles of those near home—a bit warmer or colder, or drier or wetter, but otherwise much the same. Many perpetuate the old idea that "caves in anything but limestone" have little to offer.

It just isn't so. It has been my privilege to travel fairly widely, and to cave in richly diverse areas. Yet each time I travel I am amazed anew, and the variety of rocks concealing challenging caves, and the variety of their origins, development, and features can hardly be overstated. Variety in location, content, and interaction with man. Variety in size, pattern, and delights.

Among our caves are "limestone" caves in a variety of rocks; lava tube caves; glacier caves and their strange geothermal variant, "steam" caves; littoral or "sea" caves; talus caves and fissure caves; volcanic vent caves; crusted fumaroles; piping caves; several types of caves formed in or by travertine; and a few that seem to defy classification.

A single chamber in Carlsbad Cavern is three-fourths of a mile long. In contrast, British Columbia's Slesse Creek Cave, although similarly in limestone, was so small that, before it was quarried away, happy-go-lucky northwestern cavers competed to see how many could squeeze inside simultaneously (the record was five).

Some caves are horizontal, some vertical, and some follow every intermediate angle. Floor plans vary from long, straight corridors to incredibly intricate networks and large, rounded rooms. A few are three-dimensional mazes, each passage going in its own direction, seemingly without reason. Some maintain a near-constant pattern for thousands of feet; others vary almost from foot to foot. Perhaps the simplest are merely bizarrely jumbled spaces amid fallen rocks. Some cavers feel that each cave has its own personality, and they are not far wrong.

Some caves form in ice, others form ice. Some desert caves contain large lakes and streams; some, beneath well-watered Kentucky, are bone dry. Cave waters usually approximate the average local temperature, but a few seem much too warm or cold. Many flood each year—some in winter, others in summer, depending on the onset of the local wet season. A few gush regularly, periodically, all year long, as a result of the filling and emptying of a siphon somewhere far inside.

In my own state of Washington, caves are found from sea level to the icy craters atop Mount Rainier, more than fourteen thousand feet higher. Certain types of caves and certain cave features are characteristically found in certain locations under certain circumstances. Yet the name of Surprise Pit, 437 feet deep in what seemed just an ordinary little Alabama cave, exemplifies North American caving. Here the unexpected often lies just around the next turn.

On a continental scale, the definition of "cave" must be broad: a natural underground cavity regardless of its origin, nature, or bedrock. Presumably it must be big enough to get into, and if not pitch black, pretty darned dark.

This definition is far from all inclusive. Obviously it excludes artificial underground cavities that some thoughtlessly term caves—the tunnels children dig in dirt or sandbanks with sometimes fatal "cave-ins"—as well as shallow rock shelters—rock overhangs that lack near-total darkness. (Archeologists once were wont to call these "caves," to the dismay of our systematic files.)

And especially it excludes mines. In the area where Wisconsin, Iowa, and Illinois come together, caves and dimly remembered, long-abandoned mines are closely enough interrelated that their exploration

Symbol of modern cave exploration: Harry White
"Jumaring" out of an unnamed fifty-foot Tennessee pit.

has become intertwined. Elsewhere in the United States and much of North America, cavers reject the idea of mine exploration. We venture into crumbling tunnels in search of caves accidentally intersected by long-forgotten adits. But we do so warily, for old mines are dangerous and few cavers can read their silent language, clearly a different dialect from that of caves.

The ultimate in spelean spaciousness
in the Big Room of Carlsbad Cavern.

Wherever he goes in North America, a caver can spelunk reasonably safely and happily if he anticipates variety, always viewing cave exploration as inherently dangerous—to the cave as well as to the caver. The responsible caver protects both.

The basic principles of cave exploration vary remarkably little from the summit of Mount Rainier to the Panama jungle. But precautions vary from region to region, from cave to cave. From a thousand trips into the eternal night of caves, I am deeply aware that no book can substitute for hard-won experience. Yet a book outlining the techniques, pleasures, and safeguards of modern cave exploration has been badly needed: recent years have seen too many needless caving accidents, too many of them fatal. And in less publicized tragedies, the beauty of many once-magnificent caves has been destroyed.

This is hardly new. For generations, once-splendid caves near Kentucky's Mammoth Cave have been gutted in exchange for tiny bits of the tourist dollar. An entire industry flourished around the mining of Missouri cave flowstone. A century of progressive destruction of California's once-celebrated Alabaster Cave is easily traced in accounts of visitors from half the world. In this conservation-minded generation the rate of vandalism per visitor appears much less than ever before, yet even for scientific studies of acknowledged merit, the loss of a single stalactite wrenches the caver's heart. As man counts time and his dwindling assets, caves are an endangered, unrenewable resource. Only a large, well-informed body of public opinion can preserve their myriad values. Today, few know caves and caving as they might.

Scientific study of Paradise Ice Caves system is largely accomplished in early winter.

The locales of the limestone caves of North America vary from alpine deposits of grainy marble to jungle hillsides that swallow vanishing tropical rivers.

Every caver has an obligation to halt further spelean vandalism.

Fortunately, time remains for most of our caves, but only a little time, and for fewer and fewer caves.

The National Speleological Society and smaller groups have succeeded in educating part of the public about the unique values and hazards of caves. No group or society alone can reach everyone who thinks about venturing underground. But we can do much more than we have achieved to date. This book is but another step forward.

Not every caver will agree with every conclusion and suggestion here presented. Nor can any book of this sort ever be wholly error free. But in seeking maximum accuracy, I was fortunate to have had the generous assistance and excellent advice of several of America's greatest cavers. Errors that remain are mine, not theirs. Corrections and reports of progress will always be welcome, for caving cannot afford to perpetuate error and obsolete techniques.

My choice and arrangement of topics were deliberately a trifle unorthodox. This is no encyclopedia of caving in North America, much less the other parts of the world where the challenge of caves is quite different. Instead of following a textbook outline, I have sought to weave an interpretive pattern, in the hope that it will enhance each caver's appreciation of our cherished caves.

As The Mountaineers' famous climbing text seeks to help provide its titular *Freedom of the Hills*, may this account bring a richer freedom of our beloved realm to every caver-reader.

And to all to come.

WILLIAM R. HALLIDAY, M.D.

Seattle
March, 1974

AMERICAN CAVES and CAVING

Limestone Caves and the Rocks They Inhabit

Not even Aaron Higginbotham really knew what drew him to the miserable little hole in the poison ivy patch. Perhaps he was only curious, that muggy summer day in 1810, but his wary glance showed a cool spaciousness beyond the low orifice, and then history took its appointed course. Shaping and lighting a pine torch, the lithe young surveyor slithered into the blackness beyond—and immortality among American cavers.

Through the years, Higginbotham's cave led on and on: broad airy corridors and intricate mazes unerringly threaded by whispering bat wings; miserable but challenging crawlways and narrow underground canyons leading far back under Tennessee's Cardwell Mountain; splendid natural ballrooms adorned by graceful columns, waterfalls, and icicles of stone; yawning pits concealing glassy pools inhabited by mysteriously eyeless, dead-white critters; sparkling, crystal-flecked rock piles and mammoth domes. Over and over, Aaron Higginbotham knew the exhilaration of first bringing light where light had never been.

Today a new breed of cave explorers probes the unknown with techniques unthinkable in 1810, and Higginbotham's cave still "goes." Far beyond the tourist trails that bear the commercial name Cumberland Caverns, today's explorers face the same age-old challenges, relive Higginbotham's triumphant thrills.

A generation ago, the noted geologist J Harlen Bretz removed much of the mystery from caves when he taught us how to read their silent language and opened an entire new era of scientific exploration. Yet their mystic lure remains largely unaltered—indeed, it is even enhanced —by our greater understanding. The mere word "cave" conjures rich pictures: Tom Sawyer's cave; Captain Kidd's cave, and that of many another pirate; smugglers' caves; secret refuges for fugitives from justice

—and injustice. Here are shrines linking today's busy world with forgotten pre-Columbian lore—unknown sunless depths concealing fantastic beauty and treacherous danger, homes for bats and other mysterious animals, and for our not-so-distant forebears. The age-old spell of caves is indelibly imprinted on most of mankind.

The names of our greatest are familiar indeed: Mammoth Cave; Carlsbad Cavern; Oregon Cave; Wyandotte Cave in Indiana; Wind and Jewel caves in South Dakota; Colorado's Cave of the Winds; Luray, Grand, and Endless in Virginia; Shenandoah Cavern; Cacahuamilpa and Bustamante in Mexico; Nakimu in British Columbia; and in other days, Bellamar in Cuba. Like Higginbotham's cave, these and thousands unknown to all but cavers are a single type: "limestone" caves.

Nor is this surprising, for, on any continental or worldwide scale, "limestone" caves are by far the greatest of all caverns. These are the world's longest: far-flung Mammoth Cave first, with more than 225 miles on the map, then seven more American caves currently on the list of the world's fifteen longest. These, too, are our deepest: Mexico's Li Nita–San Agustin system, currently third or fourth deepest in the world at 4,005 feet (plus a seemingly bottomless lake). These are the most impressive (the Big Room of Carlsbad Cavern, more than one-half mile long, is hard to surpass). These, too, are our most historic; like George Washington's Cave, and the Mayan caves of Yucatan, and dozens of saltpeter caves throughout the eastern United States. Limestone caves are the ones with fascinating blind fish and salamanders and near-transparent crayfish, and the most beautiful (my vote goes to Cavern of Sonora in Texas and Luray Cavern in Virginia). Indeed, "limestone caves" are the type that pose the most exciting challenges to explorer, scientist, and intellectual alike.

Even defining the term "limestone cave" is a bit of a challenge. It is easy to record that atmospheric or soil carbon dioxide acidifies subsurface water and thus produces carbonic acid, generally considered the chief agent that carves these caves. But the tiny amount of iron pyrites present in many rocks also produces significant amounts of sulfuric acid, which helps dissolve them.

Specifying them as caverns dissolved out of limestone might extend to caves in marble in California, Massachusetts, and elsewhere, for marble is merely metamorphosed limestone, and the change doesn't seem to bother the processes of cave origin. But Luray Cavern is in dolomite, chemically a rock in which much of the soluble calcium carbonate of limestone is replaced by magnesium carbonate, which dissolves much more slowly.

Remarkably similar caves near Overton, Nevada, were formed in thick beds of rock salt. Others are common in gypsum, which is not a carbonate rock and does not react significantly with carbonic acid. Moreover, caves of this sort are occasionally found in supposedly insoluble sedimentary rocks like sandstone and conglomerate, and even in some of igneous, or volcanic, origin. To complicate matters further, some caves in limestone are mere talus caves or block-creep fissure caverns, and some are wave cut or littoral (see Chapter 2).

Yet, since the features of "limestone" caverns are so characteristic that every knowledgeable caver or speleologist instantly comprehends the term, the quotation marks demanded by some purists will henceforth be omitted.

Scientific Exploration

Vast parts of North America regularly yield virgin caves to local inquiry or near-random searching. Cave hunting and cave exploration, however, are much more efficient when based on modern scientific principles. While no book can do more than record their outlines, today's skilled caver learns all he can of limestone geology and hydrology, and of the basic principles of speleogenesis—the process of origin and development of caves. Particularly important is his understanding of the geological features that control the gradient of the flow of cave forming underground waters: the bedding of the limestone and adjoining rocks, the folding, the faulting and jointing or cracking, and especially the present and past face of the land. To these he adds

Occasionally limestone cave passages follow faults.

his reading of the accessible parts of the underground world, seeing everywhere a silent language that often provides broad hints of exciting challenge, a language sufficiently clear that operators of commercial caves have sometimes been gloriously repaid for moneys spent in drilling and blasting for additional caverns suspected where no caver could force his way. And, well aware that many pages are missing from his rocky book, the scientific caver shrugs off repeated disappointments as part of the price for occasional triumph.

Some limestone caves are small, isolated structures that developed in response to strictly local speleogenetic phenomena. In some parts of the western United States where the limestone occurs in mere pods, a few occupy virtually their entire pod. Most major limestone caves, however, are segments of extensive subterranean drainage systems—natural storm sewers, if you like, although they go up and down and around and over, more like a city's pipes than its sewers. Not all are air-filled at the moment. In several great Missouri springs that are caves-to-be, water literally bursts upward under astounding hydrostatic pressure. Wakulla Springs and other less dramatic examples in Florida show vividly the probable appearance, not long ago as the geologists count time, of Mammoth Cave, and the once-mysterious Blue Holes along the coasts of the Bahama Islands and Yucatan are ordinary limestone caves recently submerged below sea level.

In contrast to the individual caves (which often vary dramatically from the overall pattern of a system of which they are a part), the extent and pattern of major limestone cave systems can often be forecast with considerable accuracy. Two groups of past and present factors are especially pertinent: (1) the volume, acidity, and temperature of water available to dissolve the rock, (2) local and regional geologic factors controlling the gradient of underground waterflow.

The Karst Process

Limestone cave systems form, enlarge, and are destroyed as an inherent part of the natural leveling of all parts of the earth. Everyone is familiar with the work of erosion on the surface. In limestone areas, the rock also dissolves downward and inward; with development and destruction of underground drainage, a unique process termed the "karst process." Characteristic of its middle stage are extensive cave systems that "pirate" much or all of the surface water.

The karstlands of central Kentucky speak eloquently in this silent language. Here a low, flat-topped series of ridges and outlying "knobs" contain many great caves besides the Mammoth Cave–Flint Ridge system, fanning out from a sweeping curve of a gorge about four hundred feet deep. Through the gorge flows the historic Green River (albeit with some difficulty, caused by the Brownsville Dam, which also seriously disturbed the lower parts of Mammoth Cave). All the subterranean waters of these ridges and karst valleys flow to the Green River. Some emerge through short segments of cave systems, more in the flood zone of the river. The slopes and body of the ridges consist of superb limestone, capped by a thin layer of other rocks, which protect the underlying limestone and its caves.

South and east of the ridges and knobs is a broad limestone plain, where the protective roof has been removed by erosion; uncountable sinkholes pock the land everywhere. Enormous volumes of water and topsoil have funneled from this Sinkhole Plain to and through the netherworld of the famous ridges and on to the Green River. Along joints and occasional faults that offered little resistance to flow, it dissolved small, then larger channels, ever uniting in a "dendritic" pattern much like that of surface streams as it followed the gentle dip of the limestone toward the river. Steep, small, zig-zag channels at the far edge of the plain speak mutely of the early stage. Where the underground channels approach the ridges, some, like Horse (Hidden River) Cave, are spacious enough to compare with Mammoth Cave itself. Long-abandoned trunk channels high above present flood stages within Mammoth Cave Ridge, Flint Ridge, and Joppa Ridge relate that this is no new process, as do numerous "crossover" and "piracy" points, where the trunk channel drainage of these huge natural conduits adjusted to the continued downcutting of the Green River. The trunk systems are the result of the rapid decline of water to a level where it can flow either gently or in tremendous seasonal floods, regularly rising and falling dozens of feet, wholly filling tubular natural sewer passages or becoming deep rivers in vaulted corridors. These are the "master caves" sought by novice and expert alike—and sometimes found.

Here and there amid these miles of Stygian passageways, the careful observer finds hints of an earlier, almost random pattern of solution of the limestone—spacious chambers with an area of more than an acre. Small networks of minor tubes that, aged and magnified enormously, make a level or tilted maze cave like Anvil Cave or Breathing Cave or South Dakota's Wind Cave. Occasionally there are traces of a crisscross, three-dimensional lacework of tight squeezeways, largely in the form of rudimentary passages along joints and blind pockets in the walls and ceiling. Some have recorded an even more primitive form of solution, honeycombed areas a bit like the inside of a giant sponge—although nowhere in the Kentucky karstland is this curious form developed as well as in the famous Boneyard of Carlsbad Cavern. The throughway drainage of these particular ridges, valleys, and plain developed too rapidly for such extensive alternate routes to form.

The dramatic central Kentucky karstlands cry out everywhere. No expertise is needed to see that the prodigious lengths of known cave are but a tiny fraction of the underground channels, nor to predict, correctly, the types of cave that are being discovered here and there: near Green River, large, near-horizontal, comparatively straight passages at various levels. Farther and farther up the Sinkhole Plain, progressively smaller, steeper, debris-choked passages are increasingly limited to a single level segmented by collapse, and often not caver sized.

Much of the key to exciting recent discoveries in these karstlands has been an understanding of a different type of water action. Separating the ridges and the knobs—actually small ridge remnants—are large and small "blind" valleys from which no surface water ever runs. Sinkholes and swallets seasonally or continuously feed water downward, much as occurs at the upper edge of the Sinkhole Plain. But here the water is not smoothly enlarging its own select route. Here it finds innumerable ancient channels never designed for these newly invading waters, and it begins to modify them to its own use—and that of cavers, blocked from ancient channels by massive ridge-side rockfall. Along the edges of the protective ridge cap, silo-like vertical shafts

The nature and sequence of cave fills often provide clues to the past history and possible extensions of limestone caves.

(usually termed "domepits," to the dismay of purists) gape down-ward. More often than not, the top of such domepits is closed, so the caver must follow many a recent stream passage—tight, twisty, jagged and muddy—before he finds himself high on the wall of a splendid natural well dozens of feet across, with many dozen yards of blackness below. If he is lucky, his domepit turns out to have been formed along a joint followed by a trunk channel somewhere below. This happens—not often, but often enough.

With the exception of these domepits, the effects of these "vadose" waters are much like those of surface waters, with which everyone is familiar. Some of their action is reminiscent of the work of a steep, cascading mountain stream, following whatever shortcuts it finds or can dissolve. Where it borrows the course of ancient sewers it incises sweeping meanders, deep channels, or canyons in their floors. Locally such streams may wash away old silt and clay, while elsewhere they may wholly plug key openings with sand, silt, boulders, and decaying plants. Dripping water also cuts narrow slots or gashes, or deposits stalactites, stalagmites, and a myriad other glassy forms of surpassing beauty. Under such circumstances, the moderately expert caver has little difficulty reading the signposts.

Especially in cave areas in steep limestone, or those with a succession of solutional features related to stream-borne cave fills, the patterns are not as easy to read. In different terrain on the opposite side of the Green River, the patterns and extent of the caves are quite different. The fantastically pitted Cockpit Country of Jamaica and other tropical karsts are marked by characteristic conical hills (called *mogotes* in Cuba, *hums* in Puerto Rico, and other local names elsewhere), perhaps a little off center where warm, moist winds blow upon them. Kentucky, California, and Montana have nothing like them. But whether in desert or alpine regions, in level, tilted, or even vertical limestones, in marble or in barely consolidated coral reefs, the language of solution is there for those who read it.

The rewards are far from certain. No one has succeeded in connecting Virginia's Butler and Breathing caves, much less in finding the missing 90 percent of similar cave passages that should comprise that great system. The Inevitable Law of Inverse Perversity predicts that someone will break through the last short gap as soon as this book goes to press. I hope so. Nothing else has worked. Yet, not too many miles away, dedicated cavers in West Virginia read the silent language and were luckier. Within a few months the Greenbriar cave system leaped onto the list of the world's longest caves—and it's still going and going and *going!*

Specific Features of Limestone Caves

The building blocks of this language group themselves into four speleological categories with preposterous but valid names: (1) speleogens, (2) petromorphs, (3) clastic fills, (4) speleothems.

SPELEOGENS

Derived from Greek root words surviving in our words "speleology" and "genesis," the term "speleogen" designates features resulting from the solution of cavernous limestone (and more, in other types of caves). At one extreme are drip slots, which resemble vertical holes drilled with jackhammers, and at the other spongework, domepits, and other large-scale examples intergrade into patterns of caves and cave systems, some of which have already been mentioned. Several ways of subdividing them are helpful: (1) abraded vs. solutional speleogens, (2) high- vs. low-velocity speleogens, (3) phreatic vs. vadose speleogens.

ABRADED VS. SOLUTIONAL SPELEOGENS

On the face of the earth, much erosion proceeds from the cutting action of water-borne rocks, sand, gravel, and even silt. High-speed floodwaters sweep large amounts of such material into and through some caves, which probably has a considerable role in the straightening and integration of throughway passages and a lesser role in the cutting of narrow canyon passages and some stream niches. Here and there rounded potholes are occasionally seen, often with a grinding rock still in place. The most obvious sign of abrasion, however, is the smoothing of irregular grooves and other patterns on soluble rocks, and the leveling of petromorphs and other insoluble matter.

HIGH- VS. LOW-VELOCITY SPELEOGENS

In addition to this locally abrasive action, high-velocity flow often produces a characteristic scalloping or fluting of the walls and floor—and of the ceiling, if the entire passage was filled. While other factors modify their patterns, these scallops are reliable indicators of the direction and the relative speed of flow: the smaller the scallops, the faster the water. While they are best observed on limestone, they develop on other rocks and even on clay fills. Each scallop is slightly off center, with its steeper slope at the upstream end. Sometimes this is difficult to evaluate, but can usually be determined by playing a flashlight at different angles along the passage.

PHREATIC VS. VADOSE SPELEOGENS

Geologically, the term "phreatic" refers to the zone of saturation below the water table where all openings are water filled, "vadose" to that above it. The terms are poorly applicable to many limestone areas where there is no true water table, but nevertheless have survived because they are useful. Technically, any solution below the surface of the shallowest cave stream should be termed "phreatic." In practice, however, such solution is grouped with that of vadose origin because its features differ dramatically from those of deep, slow-moving waters that intergrade into those which completely fill caves.

PHREATIC SPELEOGENS

Many phreatic features have already been mentioned in relation to speleogenesis and its interpretation in exploration. Detailed discussion of most is best left to technical works, but especially important to the caver are (1) spongework or honeycombing; (2) wall and ceiling pockets and joint cavities; (3) bedrock spans, pillars, and other partitions; (4) joint- and bedding-plane tubes and "anastomoses"; (5) ceiling channels.

Most of these terms are self-descriptive. Spongework or honeycombing is a complex of interconnecting rounded cavities partially separated by irregular bedrock remnants. Spans and pillars might be said to be the last stage of a dissolving network. Anastomoses are sinuous channels in vulnerable beds, never reaching caver size, the tubes straight but similarly rudimentary passages. Ceiling channels are relics of upward solution where an earlier passage has filled with sediment. Some are linear, others meander.

VADOSE SPELEOGENS

Many vadose features have also been discussed already. Among the most important are (1) vertical shafts (domepits), (2) ribbing and hackling, (3) incised meanders and grooves, (4) floor slots and stream canyons, (5) drip slots and cups, (6) potholes.

Characteristically, the walls of domepits have small, vertical ribs due to minor local variations in the rate of solution of the limestone. Similar ribs are seen in other places where films of water or rivulets descend vertically or steeply, and in the beds of streams that are free

Stream meanders and scallops (flutes) incised into marble in a California cave. Small stream cobbles seen at bottom left are not necessary for the development of these vadose speleogens.

Wind Cave, South Dakota, is unique
in its profusion of boxwork, a
petromorph.

of abrasives. Here they are sometimes termed "lamina." Some limestones instead produce an unmistakable, irregularly jagged variant
without a specific geological name but increasingly called "hackling."

Shallow streams that fill the entire width of a passage often cut
grooves into the base of one or both walls. If the limestone is nearly

horizontal, this may be difficult to distinguish from phreatic action on an especially soluble bed, but elsewhere the pattern is clear. If such a stream does not fill the entire passage, it may meander back and forth, cutting niches and sometimes half cones in the walls. If it courses virtually straight instead, its width and other factors will determine

whether a floor slot or stream canyon will result. These, however, are
sometimes mimicked by a two-stage phreatic solution with an intermediate period of sedimentation that limits later solution to the upper
part of what was once a high, narrow, water-filled cavern. Some potholes are also of phreatic origin, but arise like similar blind cavities of
walls or ceiling. These, however, are almost always filled by sediments
if any stream has later followed that passage. Stream potholes are
almost always vadose.

Speleodebaters can have a good run at the question of whether
domepits are speleogens or a special type of passage. Will White holds
the latter view—a rather heretical one. I think he's right, but I don't
know where else to list them.

PETROMORPHS

Petromorphs are secondary features of the bedrock accidentally
exposed during enlargement of the cave. Some cavers include protrud-
ing or exposed fossils within this category, but this is technically
improper. To the caver, petromorphs are primarily of interest as
indicators of nonabrasive flow or as items of scientific import, occasionally of considerable beauty. Where extensively developed, they
make notoriously jagged obstacles. Major types are (1) boxwork, (2)
clastic dikes, (3) galena and various other secondary minerals, (4)
vugs.

Boxwork and vugs are especially prominent in caves of South
Dakota's Black Hills; galena and related mineralizations, in the Iowa-
Wisconsin-Illinois corner. Oregon Cave contains an especially notable
clastic dike. Sitting Bull Crystal Cave in South Dakota, one of
America's most notable spelean spectacles, is a single, room-sized vug
containing enormous calcite crystals.

CLASTIC FILLS

Mechanical fills are often the bane of the explorer's underground
existence: breakdown, clay, mud, silt and gravel, cobbles, boulders,
quicksand, wind-borne volcanic ash, and a wide variety of trash and

garbage. Yet these often provide valuable information on the geologic history of the cave—and thus occasional leads in scientific exploration. Especially in steeply sloping caves like those in California's Sequoia area, bewildering sequences of speleogens become intelligible only when related to the sequence of fills that interrupted and modified their forms. The powdery red clay common in caves of the central United States often reveals no grit between probing fingertips; it and similar brown clays elsewhere must have been deposited under very quiet phreatic conditions. Bypasses around barriers of such material must be sought (1) where it has compacted away from the ceiling or (2) where a stream has cut into or along it.

Breakdown domes or breakdown chambers without breakdown on the floor hint at later fills that have buried the breakdown and much more passage, or—even more hopefully—of great volumes of past waters that dissolved the fallen rock and thus ought also to have dissolved major throughway passages nearby.

One special feature of clastic fills cannot be overemphasized: their extraordinary scientific values. Many fill-plugged passages contain time-covered archeological data or a sequence of ice-age animal remains. Other fills are being increasingly sought for testing and for the application of newly evolving techniques like pollen analysis, descriptive of fluctuations of local climate through the millennia. We cannot tell today what the future will demand. Cave fills are so important a chapter in the history of our continent that he who disturbs them must answer to all time to come.

SPELEOTHEMS

Once upon a time, cavers were troubled with the problem of formation (development) of cave formations like stalactites (speleothems) hanging from formations (units of bedrock) in the late stages of cave formation (speleogenesis). Hence George W. Moore invented the term "speleothem." If this were a prettier word, more people would use it—but in any case it's better than "formation"!

Speleothems are the great glory of caves, making worthwhile the untold miseries we endure, transforming drab holes in the ground to radiant, sparkling halls of graceful splendor. From time immemorial their tranquil grace has enraptured and intrigued scientist and casual visitor alike. Even the caver-tiger forgets that they may block him from great discovery or an easy way back to the surface. The acknowledged supercaver, irrevocably enmeshed in some Freudian struggle against himself, pauses enthralled in his quest for muddier mud, ever-tighter crawlways, more vicious pits to conquer. Cavers are curious (in many ways!), and much has been learned about the composition and growth processes of many speleothems. Yet the job is barely begun.

SPELEOTHEMIC MINERALS

Commonly, speleothems of North American limestone caves are formed of calcite—a mineral formed of calcium carbonate ($CaCO_3$), which may be colorless or white. Small quantities of mud, iron, or manganese minerals often turn speleothems brown, red, or black. Copper ores occasionally cause them to be blue or green.

Fairly common are speleothems composed of aragonite, chemically the same as calcite but an entirely different mineral. According to some, it should not be stable in caves at all, but it is. Others insist that it cannot be present in cold caves, but that is wrong too. (I am not, however, going to get into those hassles here!)

Speleothems of gypsum, which chemically is hydrated calcium sulfate ($CaSO_4$), are also fairly common in some scattered areas. Those of magnesium carbonate (including dolomite, huntite, and hydromagnesite) and sulfates (epsomite, mirabilite, and others), silicates (opalites), and other salts like halite and perhaps cave saltpeter (calcium nitrate) are much less common. After these comes a long list of other minerals, mostly rare and inconspicuous and not often noted by the average caver.

DRIPSTONE AND FLOWSTONE, RIMSTONE AND SHELFSTONE

Dripstone and flowstone are gravitomorphic speleothems. In case some reader wonders, they are just what the words suggest. Stalactites and stalagmites elongate in the drip line, respectively hanging and

Flowstone terraced with small gours.

standing free. Flowstone is basically a surface coating, although it sometimes hangs free for short distances—a "waterfall" of mineral, seemingly frozen in midflow. As in many speleothems, chemical and physical considerations like changes in carbon dioxide, acidity, evaporation, and agitation are factors in their deposition—matters largely beyond the scope of this book.

Usually flowstone results from broadly flowing films of water, but occasionally it is deposited in active stream courses. It commonly intergrades into stalactitic forms known by such fantastic names as "rockfish drapery."

Calcitic flowstone also intergrades into rimstone. Dolomite and hydromagnesite flowstone and occasionally calcite flowstone are found in a peculiar puttylike form called "moon milk." Its origin has been attributed to bacterial and other organic action, but proof is lacking.

Where calcite-depositing water sweeps from a crack in a wall or speleothem, a rudimentary flowstone rib may form. These are termed "welts" or "seams." "Shields" or "palettes" are wide, flat extensions of welts in which (according to some theorists) continuing flow has kept open a center strip. This does not explain their consistent magnificent roundness, but no matter. These, especially those with graceful stalactitic fringes, are among the most beautiful of subterranean sights, even though we don't understand how such forms could possibly develop.

Perhaps even more perplexing are "folia," which are thin, curved apronlike pendants. None of the various speculations about their origin seem worth including here. Although they look like variants of flowstone, this may be misleading.

Ice, gypsum, opalite, clay, and probably other forms of flowstone also exist. The first produces fine natural subterranean skating rinks.

Perched flowstone and/or dripstone deposits that originally formed atop a long-gone fill are termed "canopies," and are especially valuable clues to speleogenetic histories. The term is occasionally applied to similar features of different significance.

"Rimstone" is a comparatively nonspecific term. Where flowstone extends several inches over the surface of a pool, it forms a rimstone shelf or "shelfstone." Other shelfstone, however, extends over the surface as a result of coalescence of calcite rafts or in other ways. Thin shelfstones are sometimes mistermed "cave ice" (true cave ice is ordinary ice in caves—see Chapter 3). Where shelfstone bedecks a pool that has submerged all but the tips of stalagmites, forms resembling giant mushrooms, lily pads, or bird baths occur.

More dramatic and of different import are rimstone barriers that dam pools a fraction of an inch or many feet deep. Some of these are virtual underground replicas of hot spring terraces. They form from ice, silicates, iron, and clay minerals as well as calcite. Because the term "rimstone" is often used to include shelfstone, the French word *gour* is being increasingly used in North America, and its diminutive *microgour* for small or shallow terraces on the surface of flowstone and other speleothems. Many Mexican caves have especially magnificent terraces made up of large gours. Solitary rimstone barriers in Mammoth Cave, several feet high and dozens of feet wide, are sometimes termed "rimstone barrages" rather than merely "rimstone dams."

STALACTITES

With the exception of icicles (see Chapter 3), most stalactites of North American limestone caves are formed of calcite. The basic form is the tubular, or "soda-straw," stalactite just a little larger than a drop of water. Most are quite short, but a few reach lengths of many feet. These elongate when calcite is deposited at the ever-advancing drip

Speleothems in Carlsbad Cavern. Tapered and tubular stalactites, draperies, stalagmites, helictites, and coralloids are shown.

ring. They rarely have a crystalline surface, although during dry periods their central tube may become plugged by crystallization. Occasionally, renewed flow then causes odd nodules or bulbous, beet- or turnip-shaped stalactites, as a result of diffusion of water through the wall. Most stalactites are tapered as a result of deposition of additional calcite on the surface of the original soda straw. (Technically, this is flowstone—see page 18).

Occasionally, clumps of aragonite needles or gypsum plates assume stalactitic patterns. Masses of magnesium and sodium minerals also sometimes form tapered stalactites, and small tubular silicate stalactites are known as well.

RIBBONS, DRAPERIES, AND CURTAINS

Ribbons are a linear stalactitic form, deposited along a sloping path by slow trickles of water, usually in a straight line but occasionally sinuous. If the ceiling slopes gently, the ribbon pattern may be uniform for many feet. Large examples are often banded—hence the term "bacon rind."

With increasing slope of the roof or width of the ribbon, convolutions begin to appear at its lower tip, whereupon the terms "drapery" and "curtain" are often applied. The latter is confusingly applied also to "curtains" formed by long rows of intermingled stalactites.

Ice ribbons and related forms are fairly common in freezing caverns (see Chapter 3).

STALAGMITES

In limestone caves of North America, stalagmites are much more homogeneous than stalactites. Quite a few materials take this form, but the most commonly observed types (except ice) are calcite. Few have crystalline surfaces. The least form is an eggshell-thin lining of drip cups in mud, or small flat discs on its surface. At the opposite extreme are the familiar gigantic forms of Carlsbad Cavern, nearby New Cave, and many others dear to every caver.

When stalactite meets stalagmite the result is a column. Not a pillar,

Luray Cavern is world famous for the symmetry of its draperies and other speleothems.

Curved drapery.

which is a term applied to certain bedrock features. And certainly not a "mightytite"—perhaps the most appalling verbal creation of all time.

CORALLOID SPELEOTHEMS

The category "cave coral" includes a scientifically confused group of also confused-looking speleothems that badly need study and classification. The typical form consists of clusters of popcornlike balls or knobs at the end of thin stalks, but great variations occur. Many appear to enlarge by progressive deposition at the largest point of the knob, where evaporation and/or loss of carbon dioxide are maximum; water flow to that point is maintained by surface capillarity. Other forms appear to be flowstone-coated bunches of crystals or projecting bedrock impurities. A distinct subtype of the latter, resembling pendulous udders, is termed "mammillaries." Some large ("macronodular") coralloids are hollow, and may intergrade into thin-walled "cave blisters," some of which contain powdery gypsum or calcite. Others have tiny central canals and are probably closely related to helictites (see below). A common type consists largely of thick coats of calcite deposited by saturated cave pools. Mammillaries and certain cauliflowerlike forms also apparently form under water.

"Confused" is the proper word. But if you speak of crawling on cave coral, every spelunker sympathizes.

VERMIFORM SPELEOTHEMS

"Vermiform" means "worm-shaped," as every Latin and/or medical buff instantly notes—a good way to begin this section, for the commonest of these speleothems is the helictite, which means something like "spiraled stalactite." But, since some helictites are straight, and lots of other characteristics don't fit the image, for a group name it's really easier to figure that some worms are straight and others twist in every direction. Helictites, too, average about worm size—and those who like to carry comparisons *ad absurdum* may derive some pleasure from the fact that both have a threadlike tube down most of their length.

Most helictites are a bit thinner than tubular stalactites, but their thickness varies from about one-twentieth of an inch to two inches or more. They elongate at the tip. Nearly all have tiny central canals,

Helictites in Timpanogos Cave, Utah.

Shields (palettes) and other speleothems in Nevada's Lehman Cave.

The terms "pillar" and "column" are often confused. Pillars are vertical remnants of bedrock. Columns are fused stalactites and stalagmites. This one in the Flint Ridge section of Mammoth Cave includes several minerals but most are calcitic.

Two types of helictites in Texas' famous Caverns of Sonora.

which can be seen through the wall of the glassiest. A few have much larger canals, perhaps as a result of hydrostatic solution, and a few consist of distinct crystals. They "sprout" from welts on walls, ceilings, floors (heligmites?), and other speleothems. Somewhat similar siliceous forms in South Dakota are termed "scintillites." Aragonite produces a different sort of helictite, formed of tiny needles radiating forward and outward from an even tinier central canal. Good examples can be observed in Colorado's famous Cave of the Winds.

"Anthodite," a term frequently used and rarely defined, seems originally to have been coined for clumps of unusual helictitic speleothems ("slender, tubular branching formations") in Skyline Caverns, Virginia—longer, thinner, more granular, and more branched than the usual form. Then it seemingly jumped to calcite-coated aragonite crystals in the same cave, thence to a wide variety of mixtures of coralloids and stuff. I'd vote to return to the original usage.

At least two types of ice helictites occur in glacier caves. One is surprisingly like those formed of glassy calcite, while the other is an irregularly nodular icicle, seen especially at cave orifices, where it undergoes erratic melting and refreezing.

Oulopholites merit a separate category even in this oversimple analysis. Because of the impossible spelling, and because flowerlike clumps are common, they are usually termed "gypsum flowers." A few spiral like classical helictites, but are easily differentiated by their characteristic longitudinal grooves. Not all are gypsum; ice oulopholites sometimes sprout briefly when bitter cold suddenly strikes comparatively warm, moist cave earth, and calcite is also said to take this form. Unlike all the other members of this group, oulopholites are slowly squeezed out of tiny soil or bedrock orifices, producing small single or branching sheaves of crystalline filaments, or curving sheets an inch or yards wide.

CRYSTALS

Essentially all speleothems are crystalline, but a magnifying glass or microscope is usually necessary to prove it. Crystals large enough to attract the eye are uncommon in North American caves. Occasionally an entire stalactite has a crystalline form, and, in Virginia's Shenan-

Large oulopholites in an Arkansas
cave.

Inch-long "dogtooth spar" crystals in
a small Colorado cave.

The famous "gypsum lips" of Tennessee's Cumberland
Caverns demonstrate the extrusion of oulopholites from
cracks or pores in the bedrock.

doah Caverns and many other caves, crystalline faces sparkle fascinat-
ingly as the spelunker moves his light here and there on flowstone and
dripstone. Yet few would include such phenomena with the occasional
dramatic crystals whose sparkling beauty intoxicates the caver.

Commonest among cave crystals is "dog-tooth spar," the scaleno-
hedral form of calcite (and rarely gypsum). It occurs in small clumps
or huge masses, with individual crystals usually of "dog-tooth" size.
Exceptional examples are as much as six inches long—hippopotamus-
tooth spar, if you like. Especially in the Black Hills caves, some grow
sideways—"rice-crystal spar." Some are blunted—"nailhead spar." All
form underwater, as do pyramidal crystals, occasionally seen in upright and even inverted forms.

Needlelike aragonite crystals—and maybe dolomite, too—develop
above water in various arrangements. These acicular crystals vary from
thick mats of needles to stalactites and weirdly bunched angular
clumps of parallel bundles; tufts are especially common. To confuse
matters, some seemingly typical aragonite needles turn out to be
calcite when analyzed. Gypsum forms even larger needles, singly or as
very thin fascicles sometimes several feet long. These rarely clump like
those of the carbonate speleothems. Thickets of small, closely packed

Fascicular and radiating aragonite crystals, with some
additional speleotheic calcite.

gypsum needles are sometimes termed "gypsum grass." Epsomite also forms small needles, as do some rarer salts, including cave saltpeter (calcium nitrate).

The largest crystals found in North American caves are in the form of plates. Swordlike gypsum plates are located in caves near Naica, Chihuahua, and huge flat masses were found in cave earth in Gypsum Cave, Nevada, more famous for its coexistence of early man and ground sloth. Ethereal, foot-wide ice plates are to be found in Fossil Mountain Ice Cave, Wyoming. Calcite occasionally forms plates also, but these rarely reach a width of half an inch. Calcite plates often clump into rosettes and coralloids. Similar rosettes of small gypsum plates occasionally form large stalactitic masses—under Arkansas, some a radiant rose-pink.

FILAMENTOUS SPELEOTHEMS

In a few spelean locations, gypsum is extruded in single threadlike forms that tangle and mat into "cave cotton." Sometimes the threads are roughly parallel and loosely packed—"gypsum hair." On rare occasions they cluster into compact "gypsum rope" as much as an inch

Gypsum hair and gypsum cotton in the Flint Ridge section of Mammoth Cave.

Gypsum needles in Cumberland Caverns.

Gypsum "angel hair" and small oulopholites in
Cumberland Caverns. Fingers at lower left show scale.

thick and several times as long. The "angel's hair" of Tennessee's Cumberland Cavern is intermediate between "gypsum hair" and "gypsum grass."

UNATTACHED SPELEOTHEMS

A few speleothems develop free of any solid attachment. Calcite "rafts" float suspended on the surface of pools until their weight becomes excessive for continued support by surface tension, or until they stick to the wall and form shelfstone. "Cave bubbles" are calcified shells of bubbles resulting from splashing or outgassing at or near the surface of similar pools. The much larger "cave oranges" usually consist of a thin shell of calcite on some buoyant bit of debris.

Oolites are more or less spherical, solid accretions. They develop where an active drip of water prevents a pebble from adhering to its surroundings while it becomes more and more coated with dripstone. Pebbles are not the only nuclei—one California cave contains numerous snailshell oolites. In drip slots in mud, some oolites reach diameters of an inch or more, whereas elsewhere they usually begin to adhere when much smaller. Coarser types several inches in diameter seen in some Mexican and other caves may have a different origin.

MIXTURES AND MASSES AND MISCELLANEOUS

Cave mineralogy is far more complex than this account might suggest. Sometimes great sheets of flowstone engulf various earlier speleothems, forming a bizarre mixture. Cone-tipped spathites perplex the viewer in a few midcontinent caves. The Big Room of Carlsbad Cavern contains huge masses of granular gypsum; elsewhere it occurs as gypsum "sand," gypsum "crusts," gypsum "barrels," and just plain irregular lumps. The profusion of unusual forms in South Dakota's Jewel Cave alone would require a book-length report. And, adding a sort of insult to the glory of North American speleothems, even mud forms stalagmites.

Only rarely do speleothems serve as clues in scientific cave explorations, but to most they are its crowning glory. Those seeking expertise behind the beauty should consult the National Speleological Society's authoritative book, *Cave Minerals,* by Carol Hill.

Applying the Language of Caves—And Reading Their Hazards

The language of speleogens, petromorphs, and clastic fills provides the explorer unending hints. If a key passage is blocked, a rudimentary network close to the entrance suggests that looking for a parallel or upper-level passage may be more rewarding than digging. If the bottom of a domepit yields a tight vadose crawlway, chances of breaking out into a big cave are much better downslope than up. The vocabulary may seem endless, but the explorer need only begin to "read."

GETTING LOST

Hazards become somewhat predictable. Getting lost usually involves networks and parallel passages where the explorer can confuse one corridor with another, "losing" 90 to 180 degrees, or entering a large passage or chamber from a crawlway that may be almost undetectable on return.

Proper precautions become obvious: reliance on a compass, plus careful use of temporary markers like mountaineers' cairns or ducs (stacked rock monuments); spots of reflective tape, which are then picked up on the way out (unfortunately, pack rats sometimes do this first); small arrows smoked on rocks—preferably on small flat ones that can be turned over as you leave—pointing the way out. Plus careful observation, of course; even city streets look quite different from opposite directions at night. Mapping one's way into a new cave is the best of all ways to avoid getting lost, but I promised myself before I began this book that it would be limited to techniques of caving, not speleology. As for the proverbial ball of string, I know of one cave in which this was helpful, but this was overseas and rather ancient history—about twenty-five centuries ago, when a Greek caver named

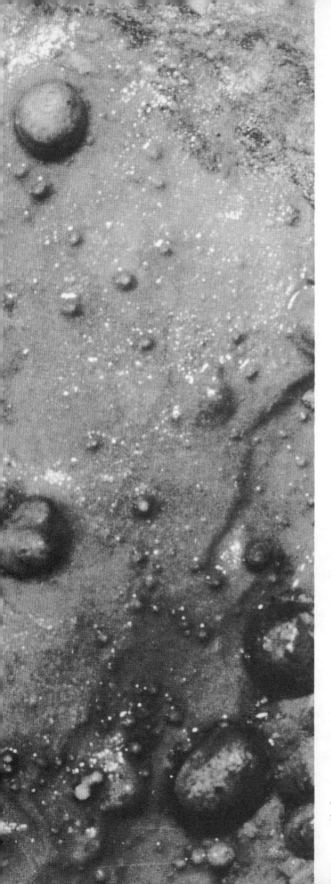

Oolites. This extraordinary group is
now on display at Carlsbad Cavern.

Some of the "balls" of Crystal Ball Cave, often proposed as a national monument.

More typical of North American caving is agonized crawling in a slurry of thin mud.

Theseus went Minotaur hunting in a Cretan limestone cave, the Labyrinth. As I've mentioned in other books, transcontinental caver George F. (Wyandotte) Jackson analyzed this particular matter years ago: if the cave is small enough for a ball of twine you can carry, it's too small to stay lost in. For those who can read the language, not even Alabama's supercomplex Anvil Cave (see Chapter 10) is an exception.

ROCKFALL

Rockfall is a little different. Anything that looks loose probably *is* loose, and the explorer should tread extra softly. Great piles of breakdown speak eloquently of past rockfalls, yet rockfall is only a minor hazard of limestone caves. Experienced cavers avoid those being drained for the first time—water-level caves at the time of Florida droughts, for example—when the most intensive rockfall occurs. All caves are in the process of falling down, but in most limestone caves this proceeds so slowly that few cavers have ever seen or heard the spontaneous fall of a single rock. Even earthquakes cause less collapse than might be expected. Only a few dozen miles from the epicenter of the prodigious New Madrid earthquakes, creaking and swaying walls scared the bejabers out of nineteenth-century saltpeter miners in Mammoth Cave, but no one saw or heard anything fall. When a tourist was killed by a rock a few years ago in a Kentucky cave, it was because he jiggled it to see if it was as loose as it looked. Not everything that looks loose is loose, but novices and old hands alike keep at least a corner of one eye cocked toward the ceiling.

Getting Around in a Cave

MUD

Actually, getting around in limestone caves really isn't greatly different from nighttime hiking, clambering around, slogging or slopping through similar terrain outdoors—in places without dense brush, poison ivy, brambles, nettles, climatic extremes, dangerous animals, and other minor surface nuisances. But mud is different.

Limestone cavers aren't required to enjoy mud, but it's a useful trait to develop. Mud sucks tennis shoes into oblivion, clumps boots into shapeless fifteen-pound masses, provides unplanned chutes, ruins cameras and tempers. In crawlways or anywhere in wall-to-wall mud, about all that can be done is to be prepared, move cautiously, and expect to wallow. Cheerful mud cavers (and I've known two or three) are the greatest.

Deep mud is a particular problem. It hides many hazards besides the giant bubbles that go *poof* with a carbide flame: sharp nails in the waterlogged timbers of extinct footbridges; rotted cardboard boxes that somehow held together long enough for broken beer bottles to be swept into the cave; layers of partially decomposed horse and cow manure; flat rocks that seem to be the passage floor until you stride off the end into the bottomless mud—the mud, mud, glorious mud immortalized by the cavers' song. It's no fun wearing boots in thigh-deep mud, but it's less fun without them. Fortunately, it is rarely necessary to crawl in deep mud of this sort.

CRAWLWAYS

Almost all types of caves require crawling, but in limestone caves it becomes routine indeed. Two well-known West Virginia cavers will never live down crawling into a patch of poison ivy. No particular problem, just crawling along unconcernedly, half asleep on their knees: (1) not realizing that night had fallen while they were in the cave, (2) not recognizing that the drops falling on them were now rain, and (3) not noticing that they had crawled halfway down the hillside from the long, low entranceway.

Some cavers assert that their companions are present solely to pull them out when they get stuck, for only by trying a crawlway out for size can one be sure if it "goes." Pockets are emptied, belts come off—even under coveralls. Helmets are pushed ahead, packs looped on a trailing ankle. Clothing r-r-rips, or comes off for another try. Cave girls suddenly become moderately less rounded. Preposterous noises echo back muffled struggles. Great sighs reveal maximum forced exhalation—anything for another quarter inch of compressibility.

Such caver compressibility reaches its effective maximum in very low, very short crawlways with plenty of width to the side. The limiting dimension varies from caver to caver—head, shoulders, chest, or pelvis (the last can occasionally be eased through by tilting). In the tightest holes, the biggest end goes first—if there is no danger of getting stuck head down.

Cavers with their maximum thickness at chest level face a special risk. By forcing out their breath, some adult cavers can squeeze through a short space only five or six inches high. Such a compression of the chest, however, can be tolerated only for the length of a single long breath—a very few inches at the most, for getting wedged by the chest with the breath out can be fatal within minutes, or even seconds. When this threatens, the victim's fellow cavers must not be fooled by gasping, vain attempts at rapid, shallow breathing. If the stuck caver cannot relax and take at least an occasional reasonably deep breath, he must be dragged out immediately—regardless of clothes, skin, and appendages.

The simplest crawlway pattern is a round tube, a shape, unfortunately, rarely matched by spelunkers. Here the shoulders are likely to be the worst problem. By extending one arm ahead, some broad-shouldered cavers become considerably rounder where it matters. Testing "before need" by crawling through wire coathangers of various sizes, with the arms in different positions, provides useful guidance.

Actually cave crawling is not true crawling at all. Rather, it includes any possible or impossible combination of wiggling, wriggling, squirming, pushing, and/or pulling with the hands, feet, shoulders, buttocks, toes, ankles, taking due or undue advantage of every infinitesimal protuberance and hollow on the ceiling, floor, and walls. Often hundreds of tiny heaves gain only a few feet—or none at all.

Studying the crawl in advance helps, but traction points are often felt rather than seen. Overlooked points on the ceiling are often the best, and, especially in tight stream meanders, frequent rotations may be needed. I'll swear that once in Flint Ridge my face, buttocks, and toes all headed in the same direction. In such contorted torture tubes, it is always well to recall the happy thought: either you'll soon be ready to be pulled out, or else the crawlway will be large enough for you to rise with a sigh of relief. At least to your toes and elbows or forearms, and maybe even to hands and knees—luxury indeed.

Except that you will probably have to crawl back the way you came.

Other crawlway problems are usually more easily met. If a hollow booming echo hints that there might be a big room ahead, it might be entered by a vertical drop. Better go head first and see. Crawling feet first, feeling for—well, anything—is more difficult, but is sometimes good life insurance if the passage is getting smaller and smaller, tighter and tighter.

In really tight oval crawlways, it makes little difference whether the caver's long diameter is horizontal, vertical, or tilted; gravity is dwarfed by the friction problem. Tilted, slightly larger crawls can be especially bad, because gravity tries to convert the crawler into a compressed mass along the lower side. In irregularly rounded tubes, sliding downward into the narrow end may be fatal. The caver's version of the torture cage where the victim could neither sit, stand, nor lie, however, is the tight T-shaped crawlway that doesn't quite allow a leg downward.

SQUEEZEWAYS

Squeezeways are essentially wide, low crawlways turned on their sides. Some are best passed by "crawling sideways," feet down as much as possible, avoiding sliding downward and bunching up. Squeezing or crawling through breakdown (talus) has problems and hazards all its own. The shape and size of the route and the looseness of the rock vary from squirm to squirm. Wedged in place by the impact of its fall, fresh breakdown is often safer than older piles in the process of settling and shifting.

CLASTIC BARRIERS

When talus crawling fails or is too risky, the question of digging may arise. Such digging runs the risk of dislodging natural keystones or additional sections of loosened ceiling. Long-handled tools, and sometimes careful timbering supervised by mine rescue experts, are essential—but at this stage of development of North American spelunking it usually isn't worth it.

Leadership

Each year, hundreds of novices visit caves, led by a single experienced caver. Here, reactions to emergencies are occasionally critical. Immediate, unquestioning responses to clear, decisive orders by the appointed leader may be essential for survival.

An observant superbat studying the antics of cavers, however, might well be puzzled at the seeming lack of leadership in many caving parties. This is no conundrum, for leadership comes in many varieties besides the imposed model. Some form of leadership is inherent in every moment of caving. Among experts competently assuming familiar roles, it is a shared leadership, nonetheless present despite its seeming absence. Even when an emergency confronts a group that has had a mere hour to learn about caving, leadership is more likely to be assumed than imposed, and the quiet competence of the old hand will add confidence on the part of the novices to their original respect. Too, he is likely to have promoted the emergence of new leadership by encouraging the brightest and most enthusiastic to check out side passages and the like: delegated leadership.

Undefinable underground as everywhere else, leadership in caving is blended from self-confidence and the earned confidence of others. Maturity and judgment derive from physical fitness, perseverance, ability, and eagerness and opportunity to learn. The best spelean leaders are as marked by quiet awareness of their ability and skill as by their willingness to step forward wherever they are needed. Yet humility is the factor that separates the true leader from the cocksure—an unassuming recognition of his human imperfection, and thus the ability and need to add further to his competence.

Digging for Caves

Much of North America's underground is still in the initial stages of near-random exploration. Here and there, cave digging—the next step

Banded marble in the stream passage
of Lilburn Cave, California's largest.

beyond scientific exploration—is under way. The number of cavers
engaged in this thankless and dangerous task, however, is so small,
and the British caving literature thereon so excellent, that those interested should refer to the section on suggested additional reading
(pages 313–322).

Stalactites, stalagmites, and flowstone obstructions sometimes form
infuriating barriers across a passage that obviously continues. Only
occasionally do cavers agree that such a blockade should be penetrated rather than merely admired. The ambivalent discoverers of such
a barrier do well to back off, take numerous photographs, and look for
a way around the beautiful nuisance. If none exists, no one should
attempt a breakthrough until the matter has been discussed—on a very
broad scale—with one's fellow cavers. Even the most successful vio-
lator is likely to find himself tagged as a vandal rather than a hero.

And this is hardly surprising, for we cavers regard "our" caves much
as we do our homes. We know and cherish beauty and challenge alike.
Truly we seek to take nothing but pictures (and we take the burned
flashbulbs home with us). We leave nothing but footprints, and as few
of those as possible. And underground we kill nothing but time (as
much as possible!).

Limestone caves are our most fragile, our most vulnerable, as well
as our greatest caves. The warm comradeship of most North American
caving is open only to those who share our concern, for, to protect our
beloved underground wilderness for all time, we look ahead.

Looking Forward

Additional techniques and concepts of scientific exploration are covered in later chapters, for the scientific approach is the key to extraordinary recent and future triumphs. Yet the nonscientific explorer need never be disheartened. I once found a nice little Nevada cave by ignorantly following the wrong directions and didn't realize what had happened until I compared cave maps, a year later. The Great Extension of Higginbotham's Cumberland Cavern lies where a veteran southeastern caver thought no passage would exist. Man knows its angel's hair, gypsum lips, incredible needle thickets, and magnificent oulopholites only because fun-loving Tank Gorin skillfully titillated Roy Davis and his other eager young friends into what he thought was a wild-goose chase. But also because Roy and the others donned their spelunking garb and went looking. Because of the very nature of limestone caves, scientific cave exploration can never wholly replace cavers' luck.

Today we glimpse the future of our continent's cave exploration perhaps as dimly as did Aaron Higginbotham the future of "his" cave when he crawled out into the sultry summer of 1810. Instantly enthralled, he settled his family nearby and lived out his span, probing ever deeper, ever conscious that each new exploration was but prologue.

Similar prologue is today's increasing tidal wave of enthusiasm and newly evolving techniques. Centuries from now, North American caving would be as unrecognizable to today's cavers as today's caving would be to Aaron Higginbotham.

Yet if we are successful in our efforts to preserve our matchless cave resources, each caver of the thirtieth century will know the same eager excitement that has lured man into the netherworld since man began.

2

Lava Tube, Glacier, and Other Types of Caves

Sightseeing through Yellowstone National Park some years ago, I spotted and joined a group of fellow stragglers from a nearby National Speleological Society convention. Bemused, they stood before the Dragon's Mouth Spring—an admirable stoopway entrance, much like that of hundreds of limestone caves throughout North America, and with only a few inches of water on the floor. But this was not quite the ordinary cave: the rock was volcanic, the atmosphere more than a little sulfurous, and every few seconds a great gush of near-boiling water boomed from deep inside.

Probably the Dragon's Mouth doesn't contain enough penetrable cave to warrant the time we spent wistfully gazing, thinking, joking, inventing, and rejecting. Yet the world's largest caves may not be limestone.

Expanding westward from the eastern seaboard, spelunking and speleology grew up together in the United States, penetrating virtually every major limestone cave area of the entire nation before coming of age. The vague influence of earlier European speleology further reinforced the concept that caves were limestone, and that holes in other kinds of rocks didn't amount to much. In Canada and Mexico, the pattern was not greatly dissimilar.

Yet in one small part of a mountain range in the state of Washington we have mapped some hundred thousand feet of lava tube caves and more than fifty thousand feet of glacier cave, and are still working at it. Two decades of systematic effort in areas largely leapfrogged by the first westward wave of speleology have brought vulcanospeleology and glaciospeleology into focus as distinct, important fields of exploration and study.

Vulcanospeleology and Lava Tube Caves

Lava tube caverns are found from the Pribilof Islands of the chill Bering Sea to recent lava flows in Latin America. These are the principal type of cave in Hawaii, in much of the northwestern United States, and in a few smaller areas. Increasingly, it appears that much of the island of Hawaii was built from lava flows which solidified into lava tube caverns. Surprisingly resembling limestone caves in certain ways, the origin, pattern, and features of these caves are sufficiently varied and significant that some cavers have come to consider them the most important type.

Even more than limestone caverns, lava tube caves are abandoned "natural sewers"—in this case, the abandoned conduits of geologically recent flows of a fluid basaltic lava known by the Hawaiian term of *pahoehoe* (pronounced *pah-hoey-hoey*). Many are small and inconsequential to all but the most dedicated vulcanospeleologist. Contrary to fast-dying legend, however, many are large, extensive, and complex. At present, the world's longest is in Hawaii: Kazumura Cave, with more than seven and a half miles mapped. One in Washington State has a width of one hundred feet. Heights of thirty or forty feet or even more are encountered, as well as pits, mazes, streams, lakes, cave-adapted animals, and many other cave features the limestone caver (dare I call him a calcareospeleologist?) has long considered his own private pride.

In some of these newly legitimized caverns, two or more levels are seen, precisely superimposed or slightly offset, as one would expect from the effects of a meandering stream of molten lava. Some of these stacked tubes remain separate for only a few yards before they coalesce, others for hundreds or thousands of feet. Occasional three-dimensional complexes and a few stacked caves with connecting pits up to fifty feet deep add excitement to their exploration. Occasional invader flows of red or orange lava provide a welcome variety, for even the vulcanospeleologist must admit that these black- or gray-walled caves are comparatively drab and gloomy. This may have had something to do with the initial lack of enthusiasm by traveling eastern limestone cavers who wandered into them.

The lavafall in multilevel Dynamited
Cave, a scientific reserve in
Washington State.

Photographer at work at the
"Meatball" of Washington state's
Ape Cave. This lava ball became
wedged when molten lava flowed
through this large lava tube cave.

Some cavern passages consist largely
or entirely of vadose stream slots.
Such speleogens are not limited to
limestone and other soluble rocks. The
smaller photo shows a stream slot in
an unusual cave in volcanic tuff.

The birth of North American vulcanospeleology may well have come from the recognition that these caves, too, have a silent language that reveals their individual histories. A few possess small white or orange-red stalactites or small rimstone deposits, like those seen in their limestone analogues, which are usually formed of drip-deposited silicon dioxide. Thin gray-black stalactites of once-molten lava hang above bulkier stalagmites built up drop by congealing drop. While some postulate other origins, most of this changeless dripstone can be traced to the slumping and dripping of a peculiarly shiny, seemingly glazed form of lava that thinly coats many such tubes. Elsewhere this glaze forms spectacular patterns resembling slumped gray toffee.

Other types of lava stalactites are less common; all, however, are easy prey to vandals—an unfortunate circumstance, for, given enough centuries and perhaps a climatic change or two, the decorations of limestone caves will replace themselves, but those of lava will not.

More durable and often much more spectacular are such speleothems and speleogens as lateral coatings, grooves, ridges, and ledges, mute evidence of a complex history of flow through the tube. Often, an entire lava tongue is seen congealed in place. Occasionally these are hollow, containing a small "tube-in-tube." At abrupt pitches, such tongues produce spectacular "lavafalls." Lava balls sometimes wedged as they passed through the tubes, forming impassable plugs or overhead landmarks. Congealed ripples and splash rings are common floor patterns. Although often concealed by breakdown, which can cause the present passage to be entirely above the original tube, the history of lava tube caves can be read almost as well as that of limestone caves.

Lava caverns are among the most youthful of all geologic phenomena. Their formation and development has been photographed in Hawaii and elsewhere, and movies of surging lava and incandescent stalactites seen through gaps in their ceilings are among man's most spectacular records of natural phenomena. But, even in favorable areas, their life span is measured in thousands of years. None, for example, is known in the vast Columbia River Basin lava flows, more than a million years old. Most are younger than the last glacial period—perhaps 120 centuries, although the experts don't agree exactly. Some in Hawaii seem to be older than any on the continent itself.

Or so at least we thought until very recently, when Oregon's Mowich Creek Cave turned up, through cavers' luck. This unimpressive cave is only about four hundred feet long, but the entrance is in a stream gully, about eighty feet down, where we thought no lava tube cave could be. Now we're going to have to start looking to see what else we've missed.

Virtually all lava tube caverns are entered through sinkholes formed by collapse or absence of a section of their ceilings. But, even if the lava seen on the surfaces is not classical pahoehoe basalt, large collapse sinks in geologically recent basalt flows usually mean significant lava tube caves underneath.

Like limestone caves, a lava tube cave is usually only a short segment of a much larger system. Here, too, are interruptions by segmental collapse or by deposits of mud, sand, and other water-borne material, or by underground lakes. But another, even more annoying process is common in these caves: lava seals or lava siphons, which plug sections of the original tube. Occasionally the seal is only an inch thick, but usually it is many feet or yards.

Complexities of the early stages of these systems produce surprisingly varied features. After the initial cooling, however, the processes of cave development stop unless there is invasion by a later flow. The leveling of tube-containing lava flows is not karstic, but pseudo-karstic. The presence of sinkholes, caves, and a limited subterranean drainage resembles the karst cycle, but little more. Progressive collapse of lava tube caverns, for example, produces lava trenches, quite different from the compound sinks of limestone.

Occasionally an entire lava tube system may consist of a single "throughway" tube descending from a vent to the point where a flow ponded or ran out of raw material. More commonly, each system has the basic form of a distributory complex with lateral branches, which form effluent passages somewhere along its length—especially at the lower end. Some effluent passages rejoin the main tube, forming confluent passages and braided systems. Many side passages are largely or entirely blocked by lava seals, but a few contain many thousand feet of penetrable complex. At the lower ends of these caves, odd, bulbous "toe cavities" are sometimes seen.

Much of these systems is often penetrable by man. In the braided Headquarters system of Lava Beds National Monument, California, more than five miles of caves and some half mile of large lava trenches are open to the public beneath about one-half square mile of flow. Here the casual visitor happily pops in and out of seemingly innumerable cavernous segments of system, connected by collapse pits and trenches. In contrast, Washington's Dead Horse Cave—a similar but more intricate complex with smaller individual passages—has only two tiny orifices despite more than a mile of mapped tubes.

LAVA TUBES AND PITS

Small lava tubes are also known deep inside at least one major fissure cave in lava (Crystal Ice Caves, Idaho). Additional factors are obviously present in the origin of such caves, and much remains to be learned. Equally rare are lava tube caves that are entered through a volcanic chimney, cone, or other vent. Such entrance pits may be more than a hundred feet deep. Unfortunately, none has yet led to an extensive cave.

THE EXPLORATION OF LAVA TUBE CAVERNS

The exploration of lava tube caverns can begin as soon as the lava ceases to give off lethal gases—even when unbearably hot for more than a momentary visit. Staff members at Hawaii National Park have slept in new-formed tubes, choosing those which seemed just the right temperature. Frank Hjort adds that they always guess wrong, and have to get up in the middle of the night and move to a cooler one.

At this early stage, the principal hazards of exploration are the giving off of noxious fumes and the thinness of crusts atop plastic lava. For scientific observation, it should be possible to enter even earlier, wearing firemen's asbestos suits and small, portable air tanks. The belay line would have to be wire rope with some sort of a mechanical instantaneous recall. Frankly, I'm going to let somebody else work out the details of this particular project and stick to those caverns that are long cooled.

The exploration of cooled-off lava tubes requires few specialized techniques, but several potential hazards emphasize their divergence from limestone caves. In cool and even temperate areas, near-invisible glare ice often lurks just inside the entrance and at the lower ends of cold systems. These caves trap much more cold air than do their calcareous analogues (see Chapter 3). The rock is perhaps the crumbliest and least stable encountered in any major type of cave, and extra care must be taken to avoid brushing the ceiling or disturbing talus. These are also the most jagged of caves, so extrarugged clothing, knee pads, gloves, and boots are highly desirable. Crawling on tiptoes with an "unbreakable" flashlight held in one hand works better than it sounds. And since these are much darker than most other caves, extrapowerful lights (and flashbulbs!) are desirable. Unpredictable compass deviations are sometimes confusing. Both foresights and backsights should be taken, and even these sometimes lead to serious error. Furthermore, the drier wilderness-area caves are often the chosen homes of various animals, some of them quite large and distinctly unfriendly. Special problems resulting from local surface conditions are evident in Hawaii, where tree roots sometimes form impenetrable barriers. Beneath pineapple plantations on the island of Maui, the spelunker may find himself wading waist deep in a morass of natural molasses—not even fermented, Frank Howarth reports.

A few lava tube caves show evidence of rare but overwhelming invasions by torrential streams—flash floods that have left thick deposits of water-borne volcanic ash, and eroded earlier sandbanks. Such catastrophes probably occur every few hundred years at most, but beware heavy thundershowers or a rain of hot volcanic ash on a thick snowpack!

It is no more than common sense to postpone lava tubing when volcanic activity fumes nearby. But if you don't, and someday in some remote lava tube cave you hear a growing rumble like a freight train and a sulfurous blast of heat chokes your suddenly parched throat as a fiery glow bursts out of the Stygian night—oh, well. Think of the publicity and remember your final consolation: Solidarity Forever!

Littoral caves form readily in lava, and other types are occasionally encountered, some of them oddities. British Columbia Speleo-Research has turned up what looks like a cinder tube. Horizontal tree casts in recent basalt flows may mimic lava tubes; vertical examples look a bit like domepits or lava chimneys. Occasionally, the venturer crawling along a tree cast encounters a junction with a second and rarely a third or more—silent evidence of a logjam in fluid lava. Such a maze qualifies as a genuine cave—at least in this book.

The odd material known as pillow lava contains hollow casts of a variety of objects. Washington State's Goldy's Cave is a hundred-foot tree cast in this type of rock. Nearby Rhinoceros Cave is just that: the cast of a hapless North American rhinoceros of long ago. Here the intuitive caver can crouch, bemused, where a monstrous heart once throbbed. Potentially much more exciting to the average caver, however, is the volcanic pit, already mentioned. Only two are known to open into anything much below, however, and most volcanic vents don't "go."

As in other rocks, some caves in lava are simply cracks in the bedrock—"fissure" caves or "tectonic" caves, as you like. In Oregon and California, "fissure" caves are several dozen feet deep. In Idaho some can be explored downward for several hundred feet. In addition to a cross section of the local rock, curious flow features are sometimes seen. And some in cold climates efficiently trap cold winter air, forming ice caves of great beauty (see Chapter 3).

An additional type of volcanic cave develops where silt or sandlike volcanic ash has formed a crust over part of the throat of a vent. Such "crusted fumaroles" are known in Alaska's Katmai National Monument, and probably occur elsewhere. Where the vent is broad and the crust thin, unwary visitors may make unwelcome discoveries—fortunately, quite short lived. A few other fumaroles are cavernous, and may be explored when gas flow ceases, but I am not aware of any significant discovery.

Solutional caves in dolerite are known in Australia, and ignimbrite blister caves have been found in Ethiopia, but I haven't been able to learn of any in North America. There's still much to be learned about caves in our volcanic rocks, however. Consider Fish Hatchery Cave, in my own state of Washington and described in *Caves of Washington* (Washington State Department of Natural Resources, Olympia, 1963). Except that it's not a lava tube, your guess about its origin is still as good as mine.

Glacier Caves and Glaciospeleology

Until recently, most North American cavers thought of glacier caves as being even smaller and less important than lava tube caves. Some pontificated that such caves melted out of glacier snouts each summer. Warm summer air, it was said, worked its way a little way back along seasonal subglacial streams and formed caves that were then obliterated each winter by glacier flow. Furthermore, when sizable caves did exist, torrential subglacial streams made their exploration impossible. This supposedly ended the matter.

Earlier in the century, however, summer tourists on Washington's Mount Rainier were sometimes guided into a sizable cave system in the Paradise Glacier. At times they visited a second cave in what some considered the Stevens Glacier but others called the Stevens Creek lobe of the Paradise Glacier (it was all one hunk of ice). An occasional person ventured a considerable distance inside, some even following a dry branch several hundred feet to the edge of the glacier, but no one seems to have been particularly interested. Moreover, in the 1940s the "tourist" caves at the head of the Paradise River collapsed, and the ice that contained them melted away completely.

A few years later someone discovered or rediscovered a cave in a small residual body of ice in the general area where part of the Stevens lobe of the Paradise Glacier had once been. For commercial reasons it was given the old name: Paradise Ice Caves. Summer tourists and

Glacier caves reveal features of glaciers that are difficult to study on the surface.

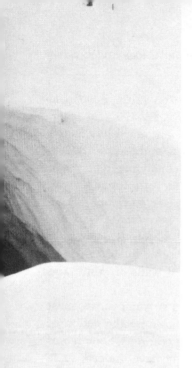

The author at one of the entrances of
The Paradise Ice Caves of Mount
Rainier, Washington.

Ablation flutes in glacier caves resemble giant stream
flutes of limestone caves.

cavers were initially able to enter only a few dozen yards. Beyond, a summer torrent filled the entire width of the cave. By 1967, however, the cave was much larger. Beneath a glacier remnant much less than a mile long, tradition-free northwestern spelunkers enthusiastically explored and mapped a complex cavern system containing more than two miles of passages. The main stream corridor was found to run completely under the little remnant glacier, tributary corridors following seasonal stream courses from its edge. Most were large, "walking" passages. Rooms were as much as 250 feet long, 90 feet wide, and 25 feet high. Glassy ice speleothems mimicking those of limestone caves multiplied the beauty of clean, nature-sculptured walls of white and blue-white ice.

Study showed that, although the cave followed stream courses, most of its size was the result of shrinkage of the glacier by the action of warm summer air coursing through the glistening corridors. Soon it was evident that this ablation was rapidly enlarging the cave. Indeed, some areas overenlarged and collapsed within a few weeks—and as some collapsed, others became passable. The entire glacier was clearly in a state of rapid retreat and shrinkage, with underground ablation playing a major role.

New techniques evolved rapidly. Many were cave-oriented adaptations of standard mountaineering techniques. Tragically, failure to follow them meticulously caused the hypothermic death of one of three cavers caught by an unexpected blizzard midway from the cave to the road and safety.

Most of these newly evolved techniques are included in later chapters of this book, but a few belong here. One is the concept of exploration in winter, when stream flow is greatly reduced. Mountaineers' ice axes are used to arrest slides on underground snowbanks, and crampons and ice creepers are used on hard snow and glare ice respectively. A unique danger of glacier caves was quickly spotted: the fall of "flakes," long, thin slabs of ice weighing a few ounces or many tons. These flakes gradually separate from the ceiling and walls and eventually peel off with a deafening crash. They remain a hazard even to the experienced glaciospeleologist because they usually remain attached at one end until the moment of collapse, and are very difficult to detect from that end. Fortunately, a loud creaking or groaning frequently precedes their fall. In other respects, the key to glacier cave exploration is survival in a cold, wet, windy environment (see Chapter 5).

Only now is international attention focusing on this new branch of speleology. Accumulating data indicates that the Paradise Ice Caves are far from unique. Glaciospeleology increasingly appears to be a tremendously promising field, long overlooked by both glaciologists and speleologists. It appears, for example, that far larger glacier caves may be present in Alaska and probably western Canada. Several major rivers burrow beneath the entire ten-mile width of Alaska's Malaspina Glacier. Just outside Juneau, dozens of square miles of the Mendenhall Glacier drain through an impressive glacier cave, oft-photographed and the subject of several picture postcards, but not yet properly investigated.

It also appears that the waters in and under other glaciers may be much more of a barrier and a hazard than at the Paradise Ice Caves. On Alaska's Martin River Glacier, moulins (glacier domepits) as much as three hundred feet deep have been seen to fill and empty within minutes when a subglacial river was blocked temporarily. And sudden drainage of intraglacial or subglacial lakes causes occasional devastating flash floods.

SNOW CAVES

With rare but dramatic exceptions, snow caves are an unimportant subdivision of glaciospeleology (it should be noted that the "snow caves" mountaineers dig for shelter are shelters, not caves at all). Large but short-lived caves are not uncommon in snowbanks in mountainous areas, their patterns much like those of glacial caves. Occasionally such a snowbank will persist for more than a year and compact into a crystalline mass known as "firn" or "névé." Heavy snowfall during the winters of 1970–1971 and 1971–1972 caused a return of more than two miles of "old" Paradise Ice Caves at the head of the Paradise River. By mid-1972 some of the firn appeared to be developing characteristics of glacier ice, and future observations here are expected to be of unusual interest. By 1978 a total of more than eight miles of passage was on the map of the combined system. By 1981 it had melted back to approximately the 1969 pattern.

GEOTHERMAL CAVES

Two large, permanent cave systems are known beneath firn, in the summit craters of Mount Rainier. In the large crater an incomplete ring of steam vents has melted a large, level cavern several thousand feet long. It extends about two-thirds of the circumference of the broad crater, approximately three hundred feet below its rim. Many side passages slope down the steep crater wall to the main corridor, and at least one leads still further downward, toward the center of the crater. Smaller caves of the same type exist nearby, and one in the lesser west crater contains a sizable lake, its temperature just above freezing.

In these geothermal caves, hissing steam vents reduce local visibility to a few inches. Their temperature varies remarkably, ranging below freezing near some of the entrances to near boiling at the vents. In these particular caves, explorers have had no problems with volcanic gases, although a strong odor of sulfur in the west crater caves hints that stays here should be short. Their exploration is hampered by the altitude of more than fourteen thousand feet, the cloudy atmosphere, and the temperature variations. The looseness of pumice slopes is a fatiguing nuisance and sometimes threatens dangerous landslides.

Not all dangerous volcanic gases have an odor, but the rule of industry applies: if the smell is detectable, there is danger. At least one person has died in such a cave near the summit of Oregon's Mount Hood. Cameras and ice axes rust after a few minutes in the geothermal caves of Mount Baker.

CREVASSES AS CAVES

Especially when covered by seasonal snow, crevasses are sometimes termed "caves." These are not truly spelean phenomena, however, unless they intersect a glacier cave. Some northwestern cavers talk

Mapping party half concealed by vapor swirling up a
steep passage of the Steam Caves atop Mount Rainier.

Water-filled pothole in Coahuila Creek Cave, a granite
talus cave in the California desert.

gleefully about chimneying crevasses from the bottom, but I don't think it's happened yet.

Glaciospeleology is even more in its infancy than vulcanospeleology. Problems of access and exploration are great, but so is the challenge. Today none can scorn glaciospeleology, for the world's largest caves may lie beneath glaciers.

Talus Caves

Especially in the western United States and Canada, other types of caves are occasionally of real interest—and talus caves are an example.

Don't sneer. Talus caves are the commonest of all caves, yet many an experienced caver has never set foot inside one. Indeed, the gung-ho caver is undoubtedly likely to scorn them as "nothing but rock piles"—which they are, for rarely do talus caves adjoin or connect important caves of other types. Yet neither the speleologist nor the sport caver can afford blandly to dismiss these not-so-simple rock piles.

In northeast and northwest America alike, intricate labyrinths dozens or hundreds of feet long writhe through seemingly impassable jumbles of rock. Unpredictably, they lead to unexpected entrances—or a blank rear wall. At least two are more than a mile long. Especially celebrated is oft-visited Bear Gulch Cave in Pinnacles National Monument, California. Here a pleasant trail remains underground for several hundred feet. In New Hampshire, other large caverns of this sort have been developed for tourists to "explore."

The commonest and simplest type of talus cave is formed by the falling, rolling, sliding, or tilting of one rock against another in a way that leaves a tiny hole or a cavernous space dozens of feet in diameter. Most are rock shelters rather than caves, but a few provide total darkness. They grade imperceptibly into a more important type, composed of a jumble of rocks of all sizes, shapes, and jaggedness, or all about the same. They may fill the bottom of a narrow canyon or gorge (a "purgatory" cave), or lie along the base of a cliff or slope. Of the myriad rock jumbles in every mountainous area, many enclose cavernous spaces that qualify as caves.

In such a cave, the explorer occasionally finds himself unexpectedly perched above a happy little stream, sparkling along in its water-polished slot, the smooth, shiny walls just right for delightful chimneying, spider fashion. It's happened to me twice in a quarter century, fifty to sixty feet beneath the searing California sun, so I've poked around in lots of drab, dusty rock piles without regret.

The origin of talus caves is lacking in mystery and usually in geologic interest. From the moment of their origin, however, they become shelters for animal and plant life that seeks the comparatively uniform cave climate. And, regardless of the type of rock and the materials on which they rest, the relentless processes of nature begin to modify them. In purgatory caves, the stream bed is usually the most interesting and sometimes the most spacious part of the cave. When they can be enticed into such a cave, geologists tend to cluster among its speleogens. Biologists, too, congregate here, but also close to the entrances—an incomprehensible trait to those sport cavers who have found rattlesnakes inhabiting the entrance zones.

In exploring a talus cave or while merely studying its features and contents, the overwhelming question is the talus itself. How solid is it?

The enormous impacts of the rocks forming talus caves ordinarily cause them initially to be invulnerable to such slight disturbances as spelunkers. As time passes, however, the rock pile settles unevenly. Beneath the rocks, the soil compacts or is washed away. Some of the rocks shift, crack, and shatter, leaving room for others to loosen. Those which look loose usually are loose. So are others that don't. Touching fractured or piled-up rock is hazardous in any type of cave, but especially in these. Often it is difficult to avoid touching the rock in tight places, but it is always a calculated risk.

These caves appear to be much less earthquake resistant than other caves, but the only eyewitness—a bear trapped six days in a Yellowstone talus cave by the great 1959 earthquake—refused to remain in the area long enough to give his National Park Service rescuers an account of the episode.

Block-Creep Caverns

Closely related to talus caves is a type of cave resulting from cracking and settling of rock parallel to the face of a cliff. These "block-creep" caverns are often further widened by frost wedging. Some are wider at the top than the bottom, others are virtually closed at the top. In time, the fissure inexorably widens, more and more rockfall occurs, and part of the long block detaches and falls. Thus they may intergrade into talus caves. Some in limestone are difficult to distinguish from small solutional caves that have undergone considerable breakdown. While most lack challenge, several fine examples are known in New York's Adirondack Mountains. In the southwestern United States, deep cavities parallel to the face of mesas trap enough cold air to form glacières (see Chapter 3).

"Sea" Caves

In North America, wave-carved caves in structurally weak zones of sea cliffs have received far less study than in Europe, where several have yielded fascinating marine life. Nor are any of our scenic littoral caves—commonly mistermed "sea" caves—even half as famous as Fingal's Cave or the Blue Grotto of Capri. Nevertheless, many such caves are known on the Atlantic Coast from New York to Labrador, and along the Pacific Coast of the United States and British Columbia. The coasts of the Caribbean and Latin America hold uncounted caves, most of them littoral. Some are found far inland, on the margins of ancient, long-dry lakes in Utah and Nevada. Many are too small to be worth recording. Others are cathedral-like vaults hundreds of feet long, worthwhile by any standard.

A wide variety of marine life inhabits these caves, including large seals ("sea lions"), the largest cave-dwelling animal of North America. Barring ecological catastrophe, these are readily observed in the impressive, commercial Oregon cave that bears their popular name. Usually peaceful and rather timid, some swim deep into such caverns to avoid visitors' boats, then panic when they run out of cave to hide in (a ton of terrified sea lion is an undesirable addition to any speleoboat-

Block-creep caves may be wider at the bottom than the top, or vice-versa.

A cavernous segment of Idaho's Great Rift National Landmark, Crystal Ice Caves are the largest known fissure caves of North America. Sections several hundred feet deep contain magnificent ice speleothems, but do not show the geological features evident here.

Not all littoral caves are on the seashore. Clinton's Cave, high above Utah's Great Salt Lake, was carved out of a cliff when the lake was fresh water and several hundred feet deeper than today.

Old postcard view of Sea Lion Cave, a very large littoral cave on the scenic Oregon coast.

Interior - Sea Lion Caves, Oregon Coast Highway 101

ing expedition). Moray eels are said to be a special hazard of some littoral caves, as are sea urchins.

While still exposed to the prodigious hydraulic "battering-ram" forces that created them, many littoral caves are unsafe to enter even when the water is at its calmest. In others, the greatest danger is the marine growth, slipperier than any cave mud. If part of the floor is bare at low tide, exploration on foot is usually easy and safe—but beware the turning of the tide! If entry is by boat or by swimming, the wave pattern is all important.

Besides Sea Lion Cave, perhaps the best-known littoral caves are Anemone Cave, Maine, and Sunny Jim Cave, largest of a group at La Jolla, California. Still larger caves are present in California's Channel Islands and along the Alaskan coast. Even hard-nosed calcareospeleologists grudgingly admit that these are real caves.

Travertine Caves

Travertine caves are locally important in scattered parts of North America. Several types exist, differing radically in their mode of origin. They have received more admiration than scientific study, and comparatively little is known about them.

Such caves are found in two dissimilar terrains: thermal "hot spring" terraces and overhanging waterfalls. Several types of the former are located in the terraces of Mammoth Hot Springs in Yellowstone National Park. McCartney's Cave, a gaping hole in the lawn in front of the park museum, is merely a funnel-shaped hot spring vent, now dried up. Several less cavernous examples are nearby. Not far away is a warm creek that runs into an elongated sink and disappears; it is either a solution cavern in travertine or natural bridging by overgrowth of an active terrace. Large cavernous examples of the latter are located in the Tonto Rim country near Pine, Arizona (Tonto Natural Bridge) and in Calaveras County, California.

Farther up the Mammoth Hot Spring terraces are the Devil's Kitchen, the Stygian Caves, and other cavernous hot spring orifices, elongated along joints rather than round, as in the case of McCartney's Cave. Swan Lakes Ice Cave is a similar example in southeastern Idaho.

Idaho's Formation Cave is a different kind of hot spring terrace cave, formed by overgrowth of a single large gour or terrace that enclosed a long, U-shaped pool.

Hot spring terrace caves often contain toxic accumulations of lethal gases. See Chapter 3 before planning to visit even the most famous.

Travertine caves also occur at the lips of certain waterfalls, past and present. These are also of several types, of which the simplest is demonstrated by Rifle Canyon Ice Cave, Colorado. This is merely an alcove in limestone; a large single column of travertine deposited by a seasonal waterfall partially walls off part of the front of the grotto. In southern California's Malibu Canyon, a broad sheet of travertine has had the same effect. Large, blunt stalactitic pendants at Idaho's Lucile Cave and Sitting Bull Falls, New Mexico, have produced much larger caves. In the latter, "ordinary" stalactites contrast curiously with blunt, porous masses of hanging travertine. Small complexes of caves have formed in similar deposits at Mooney Falls, Arizona, in Oklahoma, in another locality in Malibu Canyon, and elsewhere. In the second Malibu Canyon cave, the travertine formed as stalactites around small roots.

Piping, Stream-Cut, and Other Types of Caves

Piping (backward disintegration of a grainy rock along a water seep) and solifluction (soil flow) sometimes form caves, but these are usually minor curiosities. Officer's Cave in Oregon is a comparatively large example. Long-term studies have shown that this cave is undergoing rapid changes, so visitors should be particularly careful about roof collapse. Others are known in southern Saskatchewan, Texas, and California. Washington's Fish Hatchery Cave may be an example.

Occasionally stream-cut alcoves become cavernous, but these are rarely important to cavers. An unusually pleasant, interesting exception is Boulder Creek Cave, east of Mount Rainier in Washington State. Here a landslide dammed a canyon, forcing Boulder Creek to cut back into the cliff, where it formed a cave several hundred feet long before reaching the end of the landslide debris. The Afton

Deposition of travertine by a waterfall has formed small but pretty travertine caves at Sitting Bulls Falls, New Mexico.

Canyon Caves in San Bernardino County, California, are also of some note.

Occasionally caves and pits are found where no self-respecting cave or pit should seemingly exist. There are, for example, Mystery Hole, a large vertical pit in desert alluvium southwest of Delta, Utah; Clay Cave in Napa County, northeast of San Francisco—perhaps a "limestone solution cave" in "insoluble" volcanic tuff; the Earth Cracks of northern Arizona's Coconino Plateau, which plunge as much as five hundred feet through limestone, sandstone, and shale alike. Caves in some Lake Erie islands were probably formed by irresistible swelling of a naturally dehydrated rock. And so on. Cave exploration in North America turns up the damnedest caves, and we still have much to learn.

3

Wind and Water

"Now jist you watch," a triumphant Tom Bingham insisted. "It'll blow my hat clear down the gulch!"

But Tom's skeptical brother got the last laugh that bright South Dakota day. As Tom sailed his battered hat above the whistling little hole—*whoosht!* It vanished, sucked deep into the mysterious orifice that had been emitting a mighty wind only the day before.

History fails to record whether pioneer cowhand Tom Bingham ever got his hat back from Wind Cave—or even whether this particular tale of its discovery is any more authentic than a dozen others. Yet it could well be the true version, for Wind Cave indeed exhales prodigiously when the barometer falls, inhales with its rise. With the map of this vast cavern "growing" several miles a year, modern spelunkers follow the wind as a clue to vast volumes of yet-unknown cave.

Only a few caves—enormous caves with a single small orifice—"breathe" perceptibly with change of barometric pressure; and some seem wholly inexplicable. Indeed, as he watched his cigar smoke monotonously drift in and out of Virginia's Burnsville Saltpeter Cave every four minutes, Burton Faust momentarily questioned his own sanity. A quarter century later, all the caving world knows this as Breathing Cave, but hopeful explanations of its odd behavior still fall far short.

Most subterranean meteorology, however, can easily be observed, interpreted, and applied by the explorer. Most cavers take cave air—pure and pleasant, smog free and pollen free—for granted. Not all is odorless, tasteless; some cave air is downright nasty. But for each that is unpleasant, the aroma of dozens of cave airs hints excitingly at innate mysteries of man and nature. Crusted smoke dozens of centuries old. Bats, bears, fearless little pack rats, mushrooms, mold, and mud all leave their aromatic traces.

Spelean Temperature and Air Circulation

Unless you exercise too much or too little, somehow the temperature usually seems ideal. Relatively cool in the summer, warm in winter, they average the yearly surface temperature (unless geothermal or other exceptional factors are present). North American caves are not all the same temperature—far from it. Some are several degrees below freezing, others well above 70° F. Once well into a cave, however, the caver properly expects a remarkably even temperature whether in the Arctic or desert.

Exceptions do occur: the temperature range in the Summit Steam Caves of Mount Rainier, for instance, verges on the ridiculous. Beyond the entrance and intermediate area, however, the temperature of most caves is near constant, no matter what type of cave. Different parts of large caves may differ by a few degrees, but the uncomfortable mugginess of parts of the Flint Ridge system, for example, is due to the high level of humidity of its lower levels rather than any major temperature difference.

Hidden dangers lurk in this seeming pleasantness. In the total absence of air flow, many persons describe 40° F. air as "pleasant," and near-zero air as merely "cool." Yet air flow is rarely absent in caves. Imperceptible or gentle breezes are a major, easily overlooked cause of hypothermia (see Chapter 5).

As air and water enter the average cave, their temperatures increasingly approximate that of the local bedrock. Obviously, this does not happen immediately. Thus in winter explorations the coldest and most dangerous air currents and waterfalls are those at or just inside cavern entrances. In summer the reverse is true, with the waters and air of the inner zone of relatively constant temperature (the homeothermic

Original entrance of Wind Cave, South Dakota, where Tom Bingham is said to have lost his hat.

zone) comparatively cold, and thus more hazardous than those at the entrance.

Only in shallow rock shelters or small caves with large, multiple entrances do ordinary breezes swirl into and through caves. Instead, spelean air circulation occurs in one of several precise ways, each important alike to speleologist and caver. The simplest is wholly passive.

PASSIVE AIR CIRCULATION

As all know who have seen a balloon ascension, warm air is lighter than cold, and thus rises; cold air is heavy and settles. A few caves with a steep or vertical entrance are otherwise virtually sealed off from the surface. Such caves are potential topless cold air traps—natural deepfreezes into which cold air readily settles in winter. In summer only a limited amount of cave air is exchanged with the rest of the atmosphere, through expansion and contraction due to barometric pressure changes. If the entrance has a favorable shape, very little mixture occurs as a result of wind and rain. A few limestone caves have airtight plugs and are effective natural refrigerators, but spelean cold air traps are predominantly lava tube caves. Many such caves are colder than nearby bedrock, and some remain at or below freezing at all times. If the ceiling is high enough, the upper few feet of the cave may thaw while the floor remains frozen.

When small amounts of moisture enter such a natural deepfreeze, the result is obvious—ice. But its formation slightly warms the cave, and an excess of water will destroy the ice in turn. Consequently the balance is rather narrow. If the cave is in a region with severe winters, the stored cold may be sufficient for new-formed ice to persist throughout summer and autumn. Such caves are passive or static *glacières*—an internationally accepted French word for freezing caverns, ice caves, and other locales of subterranean ice.

ACTIVE AIR CIRCULATION

A different type of cold-trapping cave has an active air circulation resulting from air flow between entrances located at different levels. In summer the cooler, heavier air in the cave pours out at the lowest egress. To replace it, surface air is drawn into the cave through higher orifices. Usually imperceptible within broad cavern passages, such a "river" of cold air sometimes can be felt much more than a hundred yards downslope from the historic entrance of Mammoth Cave. And, before U.S. Army Engineers tragically performed explosive plastic surgery on the entrance of Virginia's Blowing Cave during World War II practice, it was famous for two hundred years for its summer air torrent.

Similar air currents are encountered at tight points in many large cavern systems but not all spelean air currents are due to this process. In some—Wind and Jewel caves, for example—reversal of the wind with changes in the surface barometric pressure reveals incredible subterranean air spaces with narrow outlets. Such barometric flow does much to equalize temperatures in far-flung parts of cavern systems.

CAVE ICE

In examples like Blowing Cave and Mammoth Cave, the direction of summer air flow reverses in winter. Comparatively warm, light air rises out of the cave and cold winter air enters its lower reaches. In small caves, the entire spelean atmosphere may be cooled below the freezing point, while in large systems inhaling great volumes of cold air, the incoming air may not reach equilibrium with the bedrock for considerable distances. Mammoth Cave sometimes sprouts transient ice stalagmites several hundred feet inside the historic entrance. In warm or temperate climates, the seasonal reversal of air flow soon begins to destroy such ice, but many caves in comparatively warm climates do develop short-lived icicles inside their entrances during cold weather.

In Arctic and high elevation caves, the average bedrock and surface temperatures may be near or below freezing. There, ice deposits in active glacières are likely to persist even longer than in pit glacières. Parts of some Rocky Mountain caves appear not to have been ice free

The exploration of ice-containing caves presents special problems.

since the last glacial period; high in the Canadian Rockies some large corridors remain wholly ice filled.

Most cave ice is seen in one of two forms. Water vapor in the atmosphere condenses into giant, often magnificent ice crystals, much as frost forms in household deepfreezes. Glare ice forms ice floors, columns, stalactites, stalagmites, draperies, helictites and other speleo-themic forms from dripping and trickling water. Travel on such ice requires special techniques.

The growth and persistence of spelean frost crystals is often dependent on a delicate meteorological balance. Despite insulating clothing, a caver's body heat may be enough to turn a glittering fairy-land of enormous, paper-thin, glassy plates into slush. Sometimes carbide lamps, gasoline lanterns, and other heating units must be used to maintain the explorers' temperature, but always at the expense of the crystalline glory of such caves. Especially in Far North and high-elevation glacières, such harm may not be undone until the next glacial period.

"Bad Air" Caves

Much has been said and written about mysterious "bad air" caves. Wet bat guano, especially that of vampires, causes "bad air" Dante could well have programed for his eighth circle of Hell. Free ammonia predominates, but other gases add their particular inimitable flavors.

More dangerous are the poisonous gases emitted by volcanic activity. Some years ago a visiting Italian speleologist ventured inside a noxious travertine cave in a sulfurous hot spring terrace near Steamboat Springs, Colorado. His breathing gear was less adequate than predicted. Properly, he was on belay—but being dragged unconscious from a cave is no fun.

Such caves are hardly typical of North American speleology, yet reports of "bad air" periodically stem from what seem to be quite ordinary limestone caves. A cave in Texas is said to contain almost pure carbon dioxide, but I cannot find any published analysis of its atmosphere. Some of the symptoms related by overburdened would-be explorers seem more typical of heatstroke than asphyxiation.

HYPOXIA

In some areas, certain passages or entire caves are said to be so deficient in oxygen that matches, candles, and carbide lamps will not burn. At low altitudes, however, most humans lose consciousness before reaching so low a level of oxygen that a carbide lamp goes out. The oxygen content of normal cave air is about 20 percent. Candles go out at about 16 percent; 15 percent is the approximate beginning level of dangerous hypoxia (lack of adequate body oxygen) with increasing, unrecognized grogginess. At 12 percent the situation is critical, and in 7 to 8 percent oxygen death is rapid. Carbide lamps are said to burn in oxygen concentrations as low as 8 to 10 percent.

CARBON DIOXIDE

Except in some geothermal caves, abnormally high levels of carbon dioxide (CO_2) cause or accompany all oxygen deficiences in North American caves. Natural processes cause CO_2 levels of 1 to 2 percent in parts of many American caves, and higher levels in a few. The human body can survive surprisingly high concentrations of CO_2. In caves it is distressing rather than lethal, and it serves as a valuable early warning signal of dangerously low levels of oxygen. Although 2 or 3 percent is about the maximum that will permit strenuous exertion, Australian "bad air cavers" have learned how to function in levels as high as 6 percent. Hard, rapid breathing in parts of Layton Cave, Missouri, has been traced to temporary CO_2 levels as high as 8 percent, the highest analysis figure I have been able to find for any "normal" American cave. This would imply an oxygen level of about 12 percent, but decreases in oxygen fortunately seem to lag slightly behind increases in CO_2.

CO_2 is heavier than ordinary cave air, and cavers' breathing may further aggravate the oxygen–CO_2 ratio—especially in deep, narrow holes. This becomes especially dangerous in rescues, particularly when the victim is wedged. In a 1977 rescue in Twiggs Cave, Maryland, oxygen readings were as low as 13 percent and CO_2 possibly as high as 18 percent. Rescuers' mental processes were drastically slowed and confused. Work normally requiring a few minutes took four times as long. Despite use of air hoses and tanks, exhaustion and vomiting were common among the rescuers. Many cases of supposed "bad air" problems seem to have involved a trace of anxiety or claustrophobia superimposed on hypothermia or heat exhaustion. But "bad air" can be a real problem. Discretion rules. Anyone developing rapid, labored breathing in any section of any cave for any reason should get out immediately (unless involved in a rescue operation as discussed later).

"Bad Air" in Caves Near Geothermal Occurrences

Travertine and geothermal caves are decisively different. Dangerous concentrations of carbon dioxide and hydrogen sulfide are often released by thermal waters and vents. When I was a boy, the National Park Service conducted tourist parties into the Devil's Kitchen, a long, vertical, slotlike travertine cave in the Mammoth Hot Spring terraces in Yellowstone National Park. Similar caves nearby were off limits because a bottom layer of carbon dioxide killed birds and small animals. Subsequent analysis of the air in the Devil's Kitchen, however, revealed that its carbon dioxide content sometimes rose above 5 percent, so it, too, is now closed to tourists and largely forgotten.

Dangerously high local concentrations of carbon dioxide have been found in at least two limestone caves that contain warm springs: Virginia's Warm River Cave and the inner rooms of Devil's Hole, Nevada. In Lower Kane Cave, Wyoming (another limestone cave), innumerable summer tourists have noted a telltale smell like rotten eggs coming from a small stream of warm water, but none has reported any ill effects. Local ranchers, however, report that in winter the stream runs dry, the smell is much stronger, and visitors have become ill. As already mentioned, the rule is that any smellable concentration of hydrogen sulfide is dangerous.

On the other hand, not all warm water in caves is of geothermal origin, and not all hydrogen sulfide is volcanic. Usually that particular odor comes from impurities in cavers' carbide. The ultimate meteorological puzzle may have been the intermittent warm spring in one Missouri cave. It was finally traced to the leaking washbowl drain of a gas station overhead.

AMMONIA

The ammonia released from wet, decomposing bat guano is normally just a nuisance rather than a hazard. The eyes and breathing apparatus alike are irritated to the extent that only an unconscious or badly injured person is likely to be exposed to hazardous amounts. Ammonia can be countered by some of the gas masks on the "surplus" market. Not all remain effective, so advance testing with household ammonia, an easy task, would be worthwhile.

INFLAMMABLE GASES AND OTHER EXPLOSIVE HAZARDS

Explosions of inflammable gases in caves are rare, yet probably more lethal than "bad air." One exception seems, at least so far, to be curious and amusing rather than harmful: exploding bubbles of inflammable gases trapped in stream-bottom mud in caves in the southeastern United States. Such gas—probably methane or a mixture of similar gases—results from the decomposition of plants swept into caves by storm waters. Momentary fireballs flare when cavers wade through such passages with carbide lights held at stream level. Large fireballs are dangerous. Cavers have been burned in West Virginia's Buckeye Creek Cave and elsewhere. Pockets of such gases might be expected where such streams end in siphons, but so far the ventilation of those caves seems to have been efficient. Potential hazards are so serious, however, that the National Cave Rescue Commission now recommends that no one use carbide lamps, candles, or any other open flame in such areas.

PETROL OR GASOLINE FUME EXPLOSIONS

At least two fatal cave explosions have been recorded in the United States. One occurred after someone dumped contaminated oil and gas in one entrance of a New Mexico cave and someone else entered another with a carbide headlamp. After the other cave explosion, a storage tank of a topside gas station was found to be leaking. Why carbide cavers entered despite the obvious smell of gasoline has never been explained. Perhaps they were cigarette smokers—a bit of a rarity among avid cavers, but a common cause of decreased smell in mankind.

GUANO EXPLOSIONS

Still another type of cave explosion belongs to a period of legend rather than speleology. It is related that in pioneer days a Texan allowed as how he would smoke a b'ar out of a local bat cave by setting fire to the guano. According to legend, the cave burned for two days after the explosion, and neither the Texan nor the bear was ever seen again (irreverent space-oriented Texas cavers snicker that they may be in orbit). Charred levels in fossil guano heaps tell of similar events long before the coming of the white man and perhaps even of the Indian. Spontaneous combustion? Perhaps, but I've decided to avoid guano caves during thunderstorms. Lightning strikes considerable distances into caves.

ACETYLENE EXPLOSIONS

When concentrated, the acetylene generated by the accidental wetting of cavers' carbide supplies is thoroughly explosive. All cavers know this, but sometimes we remember it a trifle belatedly (usually when we fail to screw the lamp bottom securely and it bursts embarrassingly into flame atop our helmets).

Alaskan caver Dave Albert is famous throughout the caving world for a serendipitous new photographic technique. It seems that, when he removed his strobe flash from his wet pack, he absently noticed that the top of his spare carbide can had come off, and replaced it—no great problem. Everything in his pack, however, was full of acetylene, including his strobe, which produced the brightest, loudest flash ever known from that type of unit. Unfortunately for the story, neither the picture nor the strobe was usable.

A commoner problem cost a sweaty northeastern caver part of his eyebrows. He was carrying a can of carbide inside a bulky rubber suit—apparently airtight as well as waterproof—and the top of the can was evidently loose. When he reached a tight crawlway the acetylene-rich air in the suit rushed out its neck. The resulting *whoom* might have been serious.

CARBON MONOXIDE AND OTHER GASES

Overwhelmingly, the truly lethal "bad air" in North American caves results from innocent, pleasant little bonfires.

Fires are safe in the mouths of rock shelters. For many years the famous Leatherman of New England slept winter and summer in campfire-warmed rock shelters along the route of his self-imposed pilgrimage. Fires are also safe in the entrances of caves that are blowing outward. If smoke is entering a cave, there is grave danger of fatal carbon monoxide accumulation. Far more persons have died this way in North American caves than from all other "bad air" combined. Little else need be said—except maybe to mention that carbon monoxide from the exhaust of a car or other gasoline- or petrol-powered equipment can be just as lethal.

Local volcanism may release gases even more dangerous than those already discussed, gases like sulfur dioxide and trioxide, which form sulfurous and sulfuric acid when they touch moist skin, membranes of the breathing apparatus, or any other moisture. Caves that smell of sulfur or ammonia are shunned by all, or should be. Yet, overwhelmingly, the airs beneath the ground are clean, bringing beauty and joy forever to the heart of the caver. Let no man pollute them.

Cave Hydrology

And there is water in caves. Rainwater.

Normally it rains during the annual conventions of the National Speleological Society. It rains and rains and *rains!* Fullgrown flash floods enlivened the convention held in bone-dry Carlsbad, New Mexico. As often as not, cars must be barred from tent-studded bogs that began as pleasant campgrounds. Perhaps no other group would accept the inevitable so cheerfully. Except from fuming vulcanospeleologists, the traditional president's greeting—"It's growing us more caves!"—draws soggy cheers.

Indeed it is growing us more caves. As should be evident from Chapter 1, rainwater is ultimately responsible for the origin, development, beauty, and much of the interest of limestone caves. It also profoundly influences cave exploration techniques throughout North America. In the Great Basin of Nevada, Lehman Cave was dissolved thousands of years ago when the desert was well watered. Today no running water can be found in the cave, yet, less than a mile distant, parts of the Baker Creek cave system flood regularly with spring freshets. A large spring downslope from Lehman Cave hints at an unknown lower level, where, when we find a way into it, explorers must beware spring floods. In the cave belt of Indiana and in the Canadian Rockies, such floods occur mostly in summer, while on Vancouver Island they come in the winter. Some tropical karst areas have two rainy seasons. Seasoned cavers inquire and plan accordingly before visiting a new area.

In California's Lilburn Cave and many another, underground water has sculptured glistening abstracts of sheer, clean beauty, and its refreshing cool is liquid delight to many a hot, tired, and dusty caver far underground. And to the caver-tiger, obsessed with the probability that a vast unknown system awaits his discovery, the coldest stream is an alluring invitation to triumph.

Yet water also means problems.

SUBTERRANEAN LAKES

Standing water holds some of the greatest delights of spelunking and some of its most exceptional miseries and frustrations. To canoe along an endless black corridor, seemingly gliding suspended in space, is incredible indeed. Echoing ripples lap at rock walls somewhere far behind, and naught else breaks the illusion of time and space suspended.

Such "canoe" caves occur in Missouri, Indiana, Tennessee, and points south, but not in the areas where speleology began in the United States. Before their existence was widely known, Dr. Oscar Hawksley of Western Missouri State College submitted a fine color slide of "speleonoeing" to the annual photographic salon of the National Speleological Society. Oz has a mercurial temper rarely matched even among cavers, and when his slide was returned, rejected as "obviously faked," American cavers were abundantly informed.

Usually, however, such waters are obstacles rather than canoeways. In tropical and desert caves, wading and swimming may be the most comfortable methods of exploration. In cold and even in temperate caves, however, they hold a serious risk of hypothermia (see Chapter 5). Other hazards exist as well, even in the tropics. Those who have unexpectedly found themselves swimming with heavy caving gear quickly learn why flotation gear is important for survival.

Decisions to swim or wade in deep water involve several other factors: the temperature of the water (and thus the risk of hypothermia), the presence or absence of mud, the weight of one's gear, the presence or absence of adequate protective clothing, and what happens if something goes wrong. Arch Cameron, one of America's outstanding cavers and manager of Mark Twain Cave, once, while neck deep in Missouri's Carroll Cave, got one foot caught between two

Some Missouri water caves like commercialized Onondaga Cave are ideal for speleonoeing.

Some caves require less luxurious craft, like this "Devil's Icebox raft."

rocks and then a cramp in the other. Fortunately he was part of a large party, which saved the day—and Arch.

For deep-water caves that aren't designed for canoes or kayaks, a variety of transportable flotation devices can be rigged. None is ideal, especially if the explorer needs to stay fairly dry. One-man or two-man rubber life rafts are collapsible, and can be dragged long distances inside Gurnee cans or other metal containers, but, unfortunately, they are heavy, easily punctured, poorly maneuverable, and seemingly always uncomfortably wet inside. Devil's Icebox rafts—named for the miserable Missouri cave where they evolved—are characteristically half a dozen small inner tubes lashed together beneath a simple flat deck. Although usually awash, they are easily portable and maneuverable in awkwardly low passages. Rubberized air mattresses serve almost as well if the caver is prepared for a bit more soaking. Still simpler and perhaps safer in narrow waters is a single inflated inner tube in which the dangling explorer paddles. They are fairly puncture resistant, and can be made more maneuverable by partial deflation at tight places, but obviously they require more efficient protective clothing. Heavy war-surplus rubber "immersion suits" combine well with inner tubes if the distance is only moderate. Unfortunately, many such suits leak, despite repeated patching. Expensive, Swedish-made floatable boaters' insulated foam suits of plasticized polyvinyl chloride have been a bit disappointing in northern waters; apparently, they have not been tested in caves.

Ping-Pong paddles make good cave paddles. To avoid losses at critical moments, they should be attached with lines that are long enough not to hinder their use. If the lake is not excessively wide, a long cord is useful (pulling the boat back avoids inefficient ferrying). Waterproof packs prized by white-water canoeists can be very useful in wet caves but, like immersion suits, must be tested and patched before need.

SIPHONS

Spelunkers are often disconcerted to find that a cave obviously continues, but underwater. This condition is known as a "siphon." The distance to an additional airfilled cave is occasionally short, and it is possible to duck under water and continue, but the mortality rate is disheartening; even experienced cavers have been known to become confused and panic fatally in easy crawlway duck-unders. Except in a few situations where the rope is likely to jam, the first man attempting an underwater duck-under should be belayed and a fixed line then emplaced. For this, stiff ropes should not be used. In the first burst of grief after a recent British cave-diving death, the belief was widespread that some vicious bystander had deliberately untied the diver's guideline. Later reflection laid the tragedy to a stiff polypropylene line, which spontaneously untied itself in later tests.

Except perhaps for a few large systems where a very short duck-under interrupts hours of tedious caving, divers' wet suits (see Chapter 3) are highly desirable for siphon use in all but the warmest caves of the United States and Canada. Some prefer to carry their clothes and a towel in a waterproof pack, passing the siphon nude—which is fine until the cavers discover that their confidence in the waterproofness of their packs was unfounded, or as long as the sudden chill doesn't cause heart attack, or until the packs get lost somewhere underwater.

CAVE DIVING

Cave diving is even more dangerous than ducking through siphons. An expert scuba diver runs little risk entering a gaping underwater cavern like Florida's spectacular Wakulla Spring, but such a channel is not typical of cave diving. In some parts of Europe, diving has become an inherent, essential tool of speleology. Far back in the watery blackness of the Mendip Hills, skilled cave divers have slowly wrenched a series of spelunking triumphs and important speleological knowledge from England's famous Wookey Hole. In North America, however, speleology has not yet marched to the point where such ventures are often essential, even though this hemisphere contains perhaps the most alluring of all cave diving areas, the fascinating Blue Holes. Just offshore in limestone areas of the West Indies and lower Latin America, these curious relics of a past when sea level lay dozens of yards below its present level were once splendid, air-filled caverns. Great volumes of tidal water now surge relentlessly through their narrow channels, and only for a few minutes at certain turnings of the tide can any diver dare these entrancing oceanic gems.

Advanced cave-diving techniques have reached a stage of considerable refinement. Special publications bridge much of the vast gap between skilled caver and skilled diver, and the National Speleological Society now has an entire Cave Diving Section, whose members have safely explored submerged caves many thousands of feet in length. In Mexico, careful coordination of cave diving and vertical caving (see Chapters 7, 8, and 9) has led to the conquest of the deepest cave currently known in the Western Hemisphere. Yet the inherent risk in cave diving is so great that I will not mention any of its techniques in this book lest they be misused. Hundreds of divers have died in waterfilled caverns of Florida alone. Most of those who died were reasonably skilled in their own element. No one should attempt cave diving until he is at home in both elements independently, prepared to deal with a maze of swimways in which visibility is reduced to an inch or two by swirls of virgin mud. The names of qualified cave diving instructors can be obtained from the National Speleological Society.

CRAWLING IN WATER

Cavers often explore streamways so tight that one's head cannot pass without removal of the helmet (the chin strap of which should be removed well in advance). In low water passages, most cavers begin by balancing on toes or knees and hands or elbows, trying to keep partly out of the water, but this is too tiring to continue for long. Usually it is much better to belly down in the water—properly dressed, of course—and let the water buoy you along. Doing just that in a miserable little cave on the Caribbean island of Barbados, Ole Sorenson dazedly found himself in the huge, beautifully decorated chambers of Harrison's Cave, currently being commercialized as a result of his discovery. In such low waterways, an occasional touch of the toes and the hands may be all the energy needed, and the use of a stick or flashlight to pole along can save some nasty hand cuts.

If the water crawl is especially tight, it may hold some unpleasant surprises. More than one caver has found himself a natural dam, the water rising toward his nose. Crawling upstream, rising water levels are so evident that serious emergencies are unlikely. Heading downstream, however, this is almost impossible to judge, and another caver must be momentarily prepared to pull the human plug. A particular danger of such explorations is panicky fear of the rising water, causing the explorer to forget to relax—often the difference between passage and disaster.

WATER CURRENTS

Except for the chill factor discussed in Chapter 5, water currents are rarely a barrier to exploration in North American caves—but dramatic

Noted geologist J Harlen Bretz and
his spelunking collie silhouetted at the
entrance of a throughway-type cave
in Missouri. Stream down-cutting and
deposition are both shown, as well as
a large block of breakdown (*left*).

exceptions do require special approaches. The raging mountain torrent of parts of British Columbia's Nakimu Caves system defies exploration except in midwinter, when even helicopter access is inconvenient. In tropical karst, deep, high-velocity streams disappear underground for thousands of feet. Subterranean cataracts, gigantic logjams, and huge, silently swirling siphons are not uncommon. Some have survived inadvertent boat rides or swims into such uncharted blackness, but more have not. By canoe or otherwise, downstream exploration of such "lost rivers" is often unfeasible, even at lowest water.

Whether upstream or down, all such explorers must be safeguarded whenever in danger of being carried free by the current or otherwise cut off from retreat. Top-quality life jackets alone are insufficient. Anchoring a length of boaters' floatable synthetic rope may permit cavers to pull themselves back upstream against a moderate current. If the length of swift water is short, the rope can be left floating until ready to return. To guard against subtly increasing currents, it should ideally be tied to the cavers' boat, with the crew venturing only a single rope length at a time until they demonstrate that they can return without the rope. In large systems, however, this isn't very practical.

As for getting soaked unexpectedly, about all that can be said is to try to avoid it, to prepare for the worst, and to prevent hypothermia if necessary (see Chapter 5). When Roger Brucker accidentally dunked himself hours inside Flint Ridge, one of his friends was packing a portable blowtorch and steamed him dry. In general, however, it is best to be prepared to get wet.

WATERFALLS

Besides the risk of hypothermia, the great risk of waterfalls is physical pounding of the body. A small waterfall may merely extinguish a carbide lamp foolishly worn instead of an electric headlight. The force of a stronger flow can numb a caver who is even fully prepared for its chill.

Descent through a waterfall is no preparation for the ascent, and the effect of a small increase in flow between descent and ascent is often not taken into consideration. If the flow doubles while the team is underground, its force increases much more than four times, often enough to sweep a climber off a ladder, or to tear the ladder apart. An increasing current may sweep rocks onto the ladder or climber. Even routes alongside waterfalls are not wholly free of such problems, for a slight increase may cause water—or rocks—to shoot far out from the lip or a ledge instead of plummeting directly.

Where waterfalls must be faced, damming and diversion may be helpful. Dams, however, have burst while cavers were in midclimb, and diversions are somewhat unpredictable. Sometimes they work fine, with the new stream course totally beyond the sight and ken of man, but more commonly it rejoins the old route, sometimes at an even more inconvenient spot.

Surface dams and diversions may also seriously inconvenience land owners and administrators, and have been known to attract thirsty cattle (it is difficult to belay attentively while being snuffled by a twitchy bull). Sandbags are much more secure than dams of dirt, rocks, sod, and other materials commonly at hand. Boards, plastic sheeting, and short lengths of pipe are often very useful, and sluice gates are highly desirable for control of dammed water. Presuming it would be neither an ecological nor scenic pollutant, the ultimate would be a permanent concrete dam with such a gate, but bags of cement are back-breaking, and at fifty-five I'm getting too old to carry my share.

When all else fails, a caver can descend and ascend through waterfalls in some form of rudimentary elevator, lowered and raised by a winch or block and tackle. Because of the enormous energy of the waterfalls where this would be attempted, all components of such systems must be able to withstand tremendous stresses. Special anchors of types discussed later may be helpful.

FLOODWATERS

In a few areas, the caves are so consistently watery that cavers soon learn to deal instinctively and confidently with the capriciousness of their stream flow. Such cavers respect the vagaries of underground waters as the mountaineer respects timberline weather. Here the experts, expecting to be trapped periodically, plan far ahead to ascertain points where they can survive. Engineers are queried about the times of release of water from reservoirs. Food, fuel, and survival gear (often including sleeping bags) are cached far into the cave.

But even the most expert sometimes misjudge the odds. In Clyde Malott's beloved Lost River cave country of Indiana, weather-wise cavers walk the plains in thundershowers, seeking geysers of turbid water two and three feet high—signs of previously unknown caverns overloaded by sudden torrents. Yet two expert Indiana cavers erred in 1961 when they entered little-known Show Farm Cave, "just for a quick look." When an unexpected ten-inch cloudburst raised the water level twenty-five feet in minutes, they correctly sought shelter at the highest point in the cave—but even that was in vain. The cave filled and caused their deaths.

Survival in such a crisis turns on judgment and experience: a split-second calculation of the chance of survival through various alternates.

The logjam at the entrance of Little Brush Creek Cave, Utah, warns cavers to beware flash floods.

Will air-filled pockets permit trapped cavers to force an entrance crawlway against a foaming tumult? How likely is it that the highest known point of a narrow nearby canyon passage will be submerged? How long could explorers survive, submerged to their necks in the flood? If all else is desperate, how much chance is there of a lifesaving discovery in yet-unknown lengths of cave?

Each case is different. The colder the water and the lower the cave, the more urgent the need for immediate escape. In tropical karst and elsewhere, if properly garbed, cavers can survive for days up to their necks in water, bobbing if necessary, wedging themselves or supporting each other to sleep. In tropical oceans, some individuals have survived for several days, floating quietly face down, barely bringing enough of the face above the water to breathe every minute or so. Cavers should be able to do the same.

Fortunately, many underground floods usually present early warning signs. Strips of rotting plant life and other debris stuck to the walls or on the ceiling provide clues to the likely water level. Rocky islands and sandbars may disappear as one gazes. The stream muddies. Floating debris increases and the chatter of the water changes, deepens ominously. Each water under the earth has its own language and perhaps its own personality.

A few of us come to cherish these dark waters as the mountaineer loves snow and ice. The rest of us merely admire their work and respect them. The winds and the waters under the earth provide much that we love. But they are deadly.

4

Headlamps

Sploop! Clank!!! With an ominous jangle, the whole front end of Reynolds Duncan's splendid new headlamp fell at his feet, its beautiful gray plastic cleanly melted.

Within days, Duncan's consternation reverberated across the continent, for this brand-new, technologically advanced carbide lamp was the last made in all North America, and in the ultramodern light that failed Reynolds Duncan—and soon a dozen others, in a dozen different ways—innumerable cavers perceived the treacherous defection of their closest, most needed, most reliable friend. But, although near-cults have long mushroomed around models or brands of headlamp to a degree that the caver's oft-desperate need for a sure source of light can only partly explain, in sober truth North American caving was more infuriated than hampered by this manufacturing fiasco. Neither the "old reliable" brass carbide headlamp or any other caving headlamp is all that it should be. Or ever has been, whether carbide-powered or electric.

Types of Headlamps

The ideal caver's headlamp is easy to describe: a rugged, reliable, lightweight, high-intensity, self-contained light, easily clipped to helmet brackets, disposable, and providing at least two hours' warm-colored light for a few cents. Nothing resembling such a headlamp exists or is likely to exist in the foreseeable future. Closest today is a tiny clip-on electric unit powered by two penlight cells, but these generate so little light, for so short a time, that they serve best as an emergency light. Yet they may be a portent of things to come.

Perhaps at the cost of some inconvenience and mild discomfort, the electric caver can go where carbide cavers cannot. He employs the extra flexibility of his chosen system to perform tasks beyond the capability of the "pure" carbide caver. The latter must turn to battery-powered headlamps when faced with waterfall pitches or air currents beyond the limit of flame protectors. Not even the most inventive, most dedicated caver has been able to find or construct a universally acceptable multipurpose electric headlamp. But with each relentless advance in electric headlamp technology, the number of carbide cavers shrinks a little more.

Nevertheless, despite all that their makers have done to discourage us, carbide headlamps remain much more popular among North American cavers.

Carbide Headlamps

A few years ago, three American manufacturers competed in providing brass "miners' lamps." All were cheap, rugged, durable, economical, efficient, and easy to use and repair.

Recognizing that some day the electric headlamp will overcome, the makers of Guy's Dropper and Autolite lamps ceased production. The Justrite Company continued production for some years, but then retooled for plastic lamps—without adequate spelean field testing. Within days of their distribution, thunderstruck cavers learned that the new gray models were dangerously brittle, difficult to use, sometimes exploded, and were vulnerable to "melt-out" of the entire front end. "Engineering myopia," shrilled the National Speleological Society *News*. A decade later, re-engineering has improved its output to the point where a few American cavers use late models consistently and effectively, but most remain skeptical.

Not surprisingly, the market boomed for the British-made Premier carbide headlamp which previously had enjoyed only sporadic North American sales (mostly in Canada). Recent competition comes from Safesport model CL-1501, a newly improved version of the Hong Kong–made Butterfly lamp previously considered too flimsy for caving. Even if production of brass models were discontinued world-wide, however, they are so long lived that they would be around for a long, long time. Long-discontinued Guy's Dropper models are not uncommon, and many a caver has a personal hoard of brass Justrites sufficient for the rest of his days.

THE PRINCIPLE OF CARBIDE LAMPS

The principle of carbide lamps is simple. The addition of water to calcium carbide (a solid chemical) produces an inflammable gas: acetylene. The burning of acetylene produces light and heat.

All carbide lamps contain compartments for water and for "carbide," as most abbreviate calcium carbide. A valve controls the water flow and thus the rate of gas production. Also present are a flint-and-steel striker to ignite the acetylene and such refinements as a gasket, a felt filter, a ceramic tip to ensure a thin, clear flame, and a surprisingly small reflector. Beautiful big seven-inch reflectors are too heavy: a clean four-inch model or even one as small as two and three-quarter inches does virtually as good a job. Many cavers add a flame protector, which helps a bit in a stiff breeze that threatens to extinguish the light. With the gray Justrite, however, flame protectors tend to cause meltoff.

Large hand-held or belt models carrying several hours' supplies are available, but most cavers use smaller, self-contained headlamps that provide two to two and a half hours' warm yellowish light per filling (considerably more if a less intense light is satisfactory).

Two basic helmet mounts are available. One has the form of a tripod; the other, a bladelike flange. Miners use both. Cavers overwhelmingly choose the flanged type because they usually stay in place during the most violent upside-down maneuvers.

Antiquers see nostalgic beauty in the clean lines of carbide lights. Few cavers concur, constantly beset by the host of minor annoyances that accompany their use. They stink a bit, and the spent carbide must be taken home (it's impure calcium hydroxide, a moderately strong caustic). Things do go wrong with them, but most of these are readily curable without special tools. Most "carbide cavers" carry a small, inexpensive lamp kit including at least a tip cleaner, a spare tip or two, and a spare felt and gasket, but some get by for years with only a tip cleaner wired to their helmet.

CARBIDE

In most of North America, carbide is available in the form of rough, slate-gray pebbles, sold in airtight cans containing from two to one hundred pounds. Large batches offer great savings to groups, but since the contents deteriorate rapidly when exposed to atmospheric moisture, empty two-pound storage cans are consequently always in demand.

All modern cans of carbide state what size material they contain, but in some remote areas very old cans containing the correct size are labeled "for bicycle lamps." A fine-grained type designed for some welding rigs is unsatisfactory for miners' lamps. Some European belt models occasionally found in Latin America require larger chunks, one-half to one inch on a side.

At least three brands of carbide—Union, National, and Shawinigan—are widely available. Union looks the prettiest, but all seem to burn equally well. Ancient cans of Radio Brand carbide made in England years ago "for bicycle and motor lamps" (sometimes found in Central America and Canada) are usually satisfactory. If the airtight seal is broken, such cans bulge and show traces of gray powder near the leak.

DISPOSAL OF CARBIDE WASTE

In the past, carbide lamps have been criticized because some thoughtless users have dumped their waste indiscriminately, even contrary to the trespass laws of some states. Not only is it an unsightly mess but, like cigarette butts, it kills small cave animals. Today's responsible cavers remove all carbide waste from caves. The easiest method is to carry one's reserve supply in spare lamp bottoms, changed every two or three hours. These extra bottoms are inexpensive, and the cost can be reduced even more by using plastic screw-on gallon jug caps instead of the metal caps sold by the manufacturer.

If no extra bottoms are available or a cap is broken or lost, a plastic bag will serve as a temporary waste receptacle. One way or another, spent carbide must be taken out of the cave. Even outside, it must not be left where curious animals can get at it, for it is a close but tasty relative of Drāno. Probably it's too weak to kill an inquisitive cow, but neither you nor the owner of said cow can be sure of that, and he's probably got a shotgun somewhere handy.

Those who simply must carry their extra carbide in something other than spare lamp bottoms often use a plastic baby's bottle—at least until they find they brought the baby's real bottle by mistake. Don't snicker. It's happened often enough for you to be next.

USE OF CARBIDE LAMPS

If you are an electric caver first confronting a brass "stinkpot," look it over thoroughly. We have found all gray plastic types very unsatisfactory, and unless it is a post-1973 model widely approved by the caving community, I would not use it. Otherwise, start at the reflector. Out toward one edge is a small flint-and-steel gadget like those of cigarette lighters. By (1) cupping your hand over the reflector after the gas has been in production a few moments, then (2) turning the wheel sharply with the palm of the hand as you sweep your hand aside, the gas ignites with a loud *plop!* (You can use a match instead, but the proper way takes only a little practice and is much more reliable underground.)

Test the wheel with a finger. If you get no spark, adjust the flint with the gadget on the back of the reflector. It may be either too tight or too loose, or the flint—and/or the tiny spring that holds it in place—may be missing. You may want to disassemble this unit to look at the flint and the spring. Be careful, for the spring is likely to jump four feet and hide.

Next, look at the center of the reflector. A nut holds the reflector onto the body. You may or may not have a wind protector—a short metal tube—clipped onto this nut. If so, remove it for a moment, so you can see the white ceramic tip (usually with a metal mount) with a tiny hole for the gas to pass.

This tip is perhaps the most important part of the lamp. Try your

Justrite, Guy's Dropper, and Autolite (*foreground*) brass
carbide headlamps. Also note unusual crystalline marble.
These spelunkers are preparing to enter California's
unique White Chief Caves, recently saved from an
ill-conceived ski development in the High Sierra.

tip cleaner to see how you will clean it when it clogs, as it will occasionally. This is rarely much of a problem, unless you weld the hole
closed by trying to clean it with the flame going.

Test the secureness of the tip. Loose tips always seem to wait until
the worst possible place to fall out. On the other hand, don't tighten it
so hard you'll need pliers to replace it hours underground; most likely
you'll find everyone left their pliers at home.

Now, unscrew the bottom compartment where the carbide lumps
go. Notice especially the rubber gasket—which sometimes falls off

when you're changing carbide in the dark. When this happens, acetylene leaks around the base and causes the lamp to burst into flame disconcertingly. Fortunately, it is easy to blow out. Carry a spare gasket—they're cheap.

With the bottom removed, you'll see a long, thin tube sticking down from the body of the lamp. Where it joins the base is an odd-shaped metal ring that holds a thin filter of felt in place. If all else seems okay and you have problems, your felt may be wet or plugged, feeling hard to the touch. These are also very cheap, and it's well to carry a spare even if you need only one a year. They last longer if they are squeezed dry only in caves, not during routine cleaning.

At the top of the lamp is a little metal bar that turns back and forth, controlling the flow of water to the carbide chamber. On the extreme left (with the lamp pointing forward as you would use it) the water is off; at the extreme right it is at maximum—far too much except for a moment to get started. With the lamp empty, run the lever back and forth a few times to see how far it goes—lamps vary considerably. Then turn it all the way off.

At this point, it is time to load the lamp. Pour a small handful of the pebbles into the bottom compartment, filling it about two-thirds full. Screw it tight. Open the small trapdoor on top of the lamp and fill about two-thirds of that compartment with water. Snap the trapdoor tightly, turn the level all the way *on* for about a second, then back about halfway. Sniff the reflector. When you smell the gas, light it with a match. If you do this too early, it will be hard to light and the flame will be small and bluish for a while. If you overdo it, you'll get a huge yellow flame—much too long—and the bottom of your lamp will be overheating. An inch of flame is about right for most conditions, for most cavers. At times you will probably want to save fuel by reducing the water supply. Some cavers get by with a half-inch flame.

Practice adjusting the flame by turning the water control lever. There is always a lag of a few seconds—much more when near the end of a load.

Next, blow the flame out and practice relighting it by cupping the palm of your hand over the reflector to trap gas, then producing a spark by a sharp motion of the heel of your hand against the striker wheel. Move your hand rapidly and completely free of the lamp or you may singe it. If you get too much gas, you'll be startled by the *bang!* If you don't trap enough gas, or you fail to get a good spark, nothing will happen. Try again.

If the bottom isn't screwed on properly, a weak flame may be a warning that a ring of fire is about to appear around the base. No great problem—it happens to me at least once a year. Just blow out both flames, turn off the water, and either tighten the bottom or remove it and reset it properly on the thread after checking to be sure the gasket is there. Doing this will dilute the acetylene to the point that it will be hard to restart the lamp for a minute or two. Don't pour in more water!

CARBIDE LAMP PARTS

Many parts of many different makes and models of metal carbide lamps are interchangeable, but some tolerances vary enough to cause problems. Premier lamp bottoms fit most Justrite lamps but not all Guy's Droppers or Autolites. Premier tips need special care in seating them firmly in some Justrite models; the same is true of Justrite tips used in Guy's Droppers. The felt holders and felt plates of earlier Premier models are quite different from those of American types. Reflector braces vary widely and often must be improvised. Justrite tip cleaners are by far the best, and its flame protector is easily modified to fit Premier lamps. Some cavers prefer homemade quarter-inch polyurethane "foam" filters to all commercial felt types. Nonmagnetic aluminum Premier reflectors are best during mapping, but most cavers prefer chromed steel Justrite types at other times.

LAMP KITS AND ANTIFREEZE

Carbide cavers' packs typically contain a soft leather pouch or film can holding the following:

Spare tips
Spare felt
Spare gasket
Tip cleaner
Spare flint

A few also carry:

Spare flint assembly
Spare tip nut ("wingnut")
Spare flame protector
Spare felt holder
Spare felt plate

but most never need these underground in a lifetime of caving.

Vodka freezes around −20° F. and, even diluted more than fifty-fifty, is an adequate headlamp antifreeze for most cave areas. On a short-term basis, it doesn't seem to change their performance, except for a need to refill more often.

BELT MODELS

Several foreign-made and domestic models of metal carbide lamps, with a large body attached at belt level and a long tube conducting acetylene gas to a helmet reflector, also are available in North America. Some of the foreign models require large chunks of carbide that are difficult to obtain in the Western Hemisphere. Most American cavers have considered them cumbersome, combining most of the disadvantages of both carbide and electric cave lights. They provide several hours' light with a single filling, however. With practice, they are easily manipulated through surprisingly tight crawlways, and their large flames are delightfully bright in big passages. Some American cavers who saw them in action during the 1981 International Congress of Speleology in Kentucky were favorably impressed.

TROUBLE-SHOOTING GUIDE TO LAMP PROBLEMS

Especially if carefully cleaned and dried after each use, a good-quality brass carbide lamp that isn't dropped too far or too often or otherwise badly mistreated will give years of faithful service. The following chart (originally devised by the Virginia Polytechnic Institute Grotto of the National Speleological Society) is a considerable

Carbide headlamp powered by
beltline case. Lava stalactites of this
size are unusual.

Dr. John F. Bridge of the Cave
Research Foundation inspecting a
poorly understood speleothem in the
Flint Ridge Section of Mammoth
Cave. Note knee pads, packs, coveralls,
chin strap, electric headlamp, and
personal touch.

help in their care and feeding. Those new to carbide lamps should not try to memorize it, but go through a few dry runs with a lamp to get an idea of what goes wrong and how easy it is to fix most of them. If you're an old hand, go on to the next section: you already know all this.

PROBLEM	INVESTIGATE	CURE
Striker assembly will not spark	Assembly for water, mud, or excessive tightness.	Adjust, clean, and/or dry.
	Flint (and spring if necessary)	Replace if missing.
No flame	Striking technique	Use a waterproof match and/or practice.
	Water valve adjustment	Increase water flow if valve is functioning.
		If broken or incorrectly assembled, replace or reseat valve. Or replace lamp.
	Carbide and water supplies, or tip for dirt, damage, or loss	Refill if necessary. Clean, or replace, seating tip solidly.
	Lamp bottom	Remove and replace, seating threads accurately and tightly.
	Gasket	Replace if worn or missing, clean if necessary.
	Felt	Replace if hard, brittle, wet, or badly worn; dry or replace if wet.
	Vent hole in water door	Clean if plugged.
	Water valve drip for stoppage	Open wide. If no better, blow through water orifice.
	Holes in lamp	Change bottoms; change lamp or patch.
Excessive flame (temporary)	Water adjustment lever	Turn off and wait for excess to clear.
(uncontrollable)	Water adjustment lever	Check seating of water valve; reseat if necessary. Replace valve assembly if broken or incorrectly assembled, or replace lamp.
Weak, angled flame	Tip for partial clogging	Clean or replace.

PROBLEM	INVESTIGATE	CURE
Flame around tip	Tip seat for damage	Try changing tip; tip and seat may have to be ground.
Water squirts out of vent hole	Water adjustment lever	Decrease flow.
	Tip for partial clogging	Clean or replace.
Bottom very hot (strong flame)	Water adjustment lever	Turn off and wait for excess to clear.
(weak flame)	Water adjustment lever and tip	(1) Clean or replace tip if partly clogged or dirty; (2) water adjustment may or may not be too high; test this, and if okay, treat as poorly burning lamp (below).
Lamp burns well but light poor	Reflector	Clean if dirty; plastic lids from coffee and other cans help protect reflectors while packed.
	Eyes	Remove dark glasses.
	Cave walls, floor, ceiling	If walls and ceiling are unusually dark, try a brighter light.
		If they are too far away to be seen and you're still underground, congratulations!
Flame around gasket	Lamp bottom	Remove bottom and replace, seating accurately and tightly.
	Gasket, threads, seat	If necessary, replace gasket and/or clean entire area. If threads are worn, replace lamp.
Burns irregularly or poorly	Tip for partial clogging	Clean or replace as needed.
	Water and carbide	Refill if necessary.
	Felt	Replace if brittle or worn, dry or replace if wet.
	Gasket and lamp bottom	Replace gasket if worn or missing, clean as needed; replace bottom seating threads tightly and accurately.
	Water valve	Open slowly. If no improvement, blow through water opening.
	Vent hole for clogging	Clean.

Electric Caving

At least a dozen commercial and innumerable homemade varieties of 3- to 12-volt electric headlamp units are currently seen in North American caves. Three types are basic: (1) "sportsmen's headlamps" and lightweight industrial models, (2) heavy-duty miners' headlamp units, (3) homemade specials.

All electric headlamps have at least one of two inevitable handicaps: the need for a wire from battery case to headpiece, and that of a considerable practical knowledge of electrical engineering for all but the simplest, least effective systems. It is easy enough to consider the energy output of batteries in terms of watt-hours, which are the product of their total voltage multiplied by the rated amp-hours. But amp-hour ratings are relatively meaningless if used with "1-amp-hour" dry cells and with low-drain bulbs. Candlepower ratings are for bare light sources, not for headlamp units. Both bulb and reflector factors affect the delivered intensity of beam and sidelight alike.

The standard measurement of battery life is the average time, under specific circumstances, at which they drop to a "cutoff voltage"—an output of 60 percent. In caving, this is especially important in NiCad and other flat-discharge batteries, which maintain a near-level output until almost exhausted. Carbon-zinc cells, on the other hand, provide useful light for many extra hours as their energy output gradually drifts downward beyond the 60 percent level. Too, the important measurement of brilliance is candlepower output rather than wattage. Under caving conditions, candlepower is especially affected by voltage changes. For example, 5 candlepower is almost identical to 5 watts, but 2 candlepower is equivalent to 3 watts. The serious electric caver must know the characteristics of a considerable variety of systems under standard and forced operation. In parts of Latin America, for example, he finds the current alien to his 60-cycle, 110-volt rechargers just when he needs a charge. Yet increasing numbers of cavers are going electric to select lighting systems suited to specific needs.

Many assert that the old bugaboo of the connecting wire is vastly overrated. Using a waist-level case on a separate belt, skilled devotees assert that they have never gotten hung up or even been inconvenienced by the wire. Such devices can be slid back and forth and around, conforming to the changing geometry of a crawlway. Running the wire inside clothing definitely hinders this action. Others try to find a better way to carry the battery case—in the pack, in shoulder harnesses, even in pockets. Most return to the belt.

Not only must the problem of getting caught by the wire be considered, but undue stress to the components of the system. Heavy, mudproof wires and rugged terminals and other construction are obviously desirable. A surprisingly successful alternate, however, avoids strains on other parts of the system: flimsy pull-apart plugs separate with minor stress and are easily plugged back together.

USE OF ELECTRIC HEADLAMPS

Because of the notable adaptability of electric caving, the expert rarely relies entirely on any single headlamp system. Instead, he chooses the power source and the bulb that will best meet his specific need. Such matters as cave temperature, reflectivity of the walls, and length of the trip are only part of the consideration. When speleohistorian Harold Meloy unexpectedly needed to study historic signatures in Mammoth Cave's Gothic Avenue, on-the-scene cavers howled with glee when he borrowed a bracketless lightweight sportsmen's headlamp, strapped it on his fedora, and nonchalantly went spelunking in his business suit. Perhaps the other extreme is encountered during photographic trips in Washington State's Dynamited Cave, a difficult lava tube cave with several vertical pitches. Not only do photographic parties function best when especially brilliant light is available, but many lava tubes have reflectivities so low that they seem to eat light. Here an Exide rechargeable MF-2 sealed lead-acid battery might well be chosen despite its bulkiness (about 5 by 4 by 3 inches). With a #965 bulb (a standard screw-base lamp with a rated life of about fifteen hours at 0.5 amp and 5 watts), this battery will yield eleven hours of 10 candlepower light (and special thanks to Bill Varnedoe, electric caver extraordinary, for this suggestion).

Several lightweight electric headlamp units are currently sold in the United States and Canada, the so-called sportsmen's headlamps. Expensive but among the best are Justrite industrial models of its 1704 series, with a case for four D-cell batteries (its 1904 series is much lighter and flimsier). Others use a bulkier 6-volt "lantern" battery. Still others have a switch that alternately connects a 3-volt bulb to pairs of D-cells. Justrite model 1903–5 is advertised to be unusually flexible, using either a 6-volt lantern battery or eight D-cells, which yield 6.3 candlepower for a few hours with a #965 bulb. Few store-bought models currently have the blade bracket considered essential for serious caving, but these are easily mounted, and ready-made models can be ordered from Justrite specifying the mount (part #20182). Justrite also manufactures refractive "honeycomb" lenses for a wide-angle light. The normal clear lens produces a spot beam.

Models without a mounting bracket can be especially useful during rescues, when only nonbracketed helmets are available for some of the noncaver manpower. The elastic headstraps supplied with these lamps are surprisingly hard to dislodge from construction-type helmets, and special hooks are available for better anchoring (Justrite part #20185).

The most vulnerable features of these lightweight units are the wire terminals, the battery contact springs (especially with mercury and other heavy batteries), and switches. The headpieces are much more reliable than the battery cases, many of which are best thrown away, saving the headpiece for construction of a reliable unit. Two- and three-inch Justrite headlamps can be ordered, as parts #1904–2 and 1903–2 respectively, without a case.

SELECTION OF DRY-CELL BATTERIES

All these lightweight units use D-cell or lantern batteries. Most rely on the former. Until recently, electric cavers had little choice in this and other common battery sizes: first-line zinc-carbon cells like the familiar Eveready and Ray-o-Vac and similar brands, or second-line zinc-carbon cells, which cost a little less but included some brands which weren't worth much. Today the situation is bewilderingly different—different sets of chemicals, different sizes, different price ranges.

Despite extra initial cost, some of the new batteries are of special value to cavers under certain circumstances. In other situations, the old reliable carbon-zinc types are at least as good.

"ORDINARY" CARBON-ZINC BATTERIES

The ordinary carbon-zinc or LeClanche dry cell has been in use for a century. Their annual sales in the United States alone approach five hundred million, and they can be bought even in remote country stores (at some risk—see below). One U.S. manufacturer lists more than a hundred different styles of carbon-zinc batteries and cells, and nearly as many different terminal arrangements. Most cavers, however, are interested only in the standard D-cell (regular flashlight), and perhaps the C-cell (thinner flashlight), AA (penlight), and AAA (slim penlight).

The National Bureau of Standards has established specifications for these and other batteries. Their chief concern to the average caver is in selection of unfamiliar brands. Those U.S. companies which advertise that their batteries meet or exceed these standards consistently manufacture and sell reliable batteries. Cheaper brands made by their overseas subsidiaries and sold in the United States are also usually reliable.

As everyone knows, such batteries are about two inches long. They are rated at 1.5 volt and sometimes produce this when new. Anything over 1.1 volt is considered satisfactory. They discharge slowly and relatively uniformly in constant use, and a considerable reserve remains when they have dimmed perceptibly. When properly cared for, most are good for at least four hours' steady use in a hand-held flashlight or headlamp using an 0.5-amp bulb of appropriate rated voltage (1.5 volt per battery). Used five minutes and rested fifteen, some are said to have twice the life of those in steady use. They are most efficient above 70° F., and their energy output is approximately halved at half that figure. Their shelf life can deteriorate as much as 20 percent per year. The destructive action of shelf storage is said to be halved for every decrease of 10° F., and is greatly accelerated by heat encountered in glove compartments, car trunks or boots, country stores, ship holds, and the like. Whenever possible they should be bought from air-conditioned stores with a rapid turnover of stock, thence immediately transferred to the refrigerator; but they should be removed at least eight hours before use. They are especially short lived when used cold for even a short time.

Some of these "ordinary" batteries seem to be slightly rechargeable, especially if only slightly discharged. But they don't charge up to their original strength, and the results don't seem to make the effort worthwhile.

PREMIUM CARBON-ZINC DRY CELLS

Several manufacturers market premium or heavy-duty D-cells. At least in some cases, the extra cost is worthwhile. Testing has shown that the Eveready premium battery has about twice the life of the excellent "ordinary" Eveready D-cell at 60 percent higher cost. An RCA premium AA battery yields 40 percent more service for 20 percent extra cost.

LIGHT-DRAIN AND HEAVY-DRAIN CARBON-ZINC BATTERIES

All carbon-zinc batteries work best in light, intermittent use. Some manufacturers chemically tailor their products for "light drain," "medium drain," and "heavy drain." The first has an especially good shelf life, and is designed for home use and for transistors. Medium-drain batteries are all-around types, perhaps better for emergency flashlights discussed in Chapter 5. The heavy-drain types stand up better under continuous use. They have a particularly short shelf life, but are often a good choice for cavers' headlamps.

"LONG" CARBON-ZINC BATTERIES

A recently marketed type of carbon-zinc dry cell is twice as long as the ordinary D-cell. It is rated at 3 volts, and is said to be more efficient than two 1.5 volt dry cells. They are better for caving than their two-cell analogues but are not widely available.

MERCURY BATTERIES

Zinc-mercury batteries yield great energy in proportion to weight and size. They have an extremely long life on the shelf or in heavy use, and a level discharge rate. D-cells are rated at 14 amp-hours. For operation above 70 or 100° F. (depending on the authority reviewed), they are the best available despite a relatively high cost. Unfortunately, they don't work very well above 0.5 amp, or in caves below about 50° F. unless taped to the skin or otherwise warmed. The Diablo Grotto of the National Speleological Society recommends operating them at 0.5 amp to keep them warm.

NICKEL-CADMIUM BATTERIES

Reliable and rugged even in difficult caving, good quality 2- to 4-amp-hour NiCad batteries are an excellent choice for most North American spelunking. These are rechargeable hundreds of times, requiring fourteen to sixteen hours in standard units sold widely throughout the continent. Despite the high initial cost of the batteries and recharger, they cost much less per hour of light than comparable zinc-carbon types. Each charge provides at least as much life as a zinc-carbon cell—somewhere around four hours with an 0.5-amp bulb. Their initial voltage is slightly lower than the rating for the zinc-carbon type (about 1.25 volts), and the candlepower of a given system is considerably lower. Their output, however, remains about 1.2 volts until the battery is almost wholly discharged. This particular characteristic can be an advantage as long as the light lasts, but a potential embarrassment for the unprepared. Bill Varnedoe points out that when they are repeatedly discharged partway, they begin to behave as if that level is their total discharge. This can be corrected, but not in a cave.

NiCads have a much greater temperature tolerance than most other batteries, operating at 60 percent capacity at −14° F. and 93 percent at 113° F. Unless they become so hot that they feel warm to the touch, their shelf life is comparatively good. For the caver who takes numerous short trips in comparatively cold caves, NiCads are hard to beat. However, they are subject to permanent damage if left long enough in a flashlight for one cell to go dead and be subjected to "reverse charging" by the others.

ALKALINE BATTERIES—RECHARGEABLE AND OTHERWISE

In caving, alkaline batteries function best in long, low-drain and short, high-drain situations. Nonrechargeable 10-amp-hour types cost two to three times as much as comparable zinc-carbon cells but last several times as long and their shelf life is better than that of zinc-carbon cells. Rechargeable types are less favorable. They cost at least twice as much as NiCad models. On ordinary rechargers, they can be reenergized only some thirty to fifty times, each time a little less than before. When well cared for, their charge retention is good. When discharged completely, however, they cannot be recharged further. This is a particular hazard because of their flat discharge curve. They function fairly well at low temperatures but not as well as NiCad or silver-zinc batteries. Some feel that alkaline D-cells should only be used with bulbs rated below 0.5 amp. For those who wish to try them but have difficulty finding a source, camera stores or electronic mail-order houses are good bets.

"PHOTOFLASH" DRY CELLS

By modifying the chemicals in zinc-carbon batteries, they can be manufactured to deliver high initial surges of current. This is valuable for flashguns, but is achieved at the expense of other characteristics desirable in caves. Except in emergencies, such batteries should be reserved for their designed purpose.

"RESERVE" BATTERIES

Magnesium-silver chloride or magnesium-cuprous chloride "reserve" or "dunk" batteries are theoretically an ideal emergency light source because of extra-long shelf life. One manufacturer claims that they will remain inert for as long as twenty years without significant energy loss. They are designed to remain "in reserve" until the cap is turned, releasing liquid that immediately activates the battery. They must be used soon after such activation, and how they will fare in twenty years' caving is an unanswered question. Newer zinc-air "reserve" batteries may prove more practical for caving but are even less tested underground.

LITHIUM CELLS AND OTHER FORMS

Lithium cells are popular among northwestern mountaineers and cavers who must contend with unusually cold caves. Other energy sources are undergoing intensive research, and some other batteries, like the silver-zinc D-cell, are in limited use. Power sources of future headlamps may well be unrecognizable as batteries as we now know them.

VARIETIES OF WET-CELL UNITS
HEAVY-DUTY RECHARGEABLE MINERS' LAMP UNITS

For many years, large mining companies have relied on rugged, reliable, rechargeable battery-powered wet-cell units designed to provide an effective light for slightly more than a normal eight-hour shift. Two basic types exist: lead-acid and nickel-iron (Edison cell). Nickel-cadmium units have never become popular. Used models of older styles are often available cheaply from British and American mines. Electric cavers have found them inexpensive and usually satisfactory.

NICAD WET-CELL UNITS

The energy output, weight, and bulk of the belt case of the new nickel-cadmium units compare favorably with the traditional types, and their price is competitive. Used models are rarely available, but the few cavers using them report that they are quite leakproof and function very well under adverse conditions. Obtaining the proper electrolyte solution (which contains traces of lithium compounds as well as potassium hydroxide) can be a problem.

Alkali leak in Nife unit of British origin.

NICKEL-IRON (EDISON OR NIFE) WET-CELL UNITS

Much commoner are rechargeable wet-cell units utilizing an alkaline electrolyte solution and nickel and iron electrodes. The American-made Edison type was the original miners' wet-cell unit, and British "Nife" models are especially popular in Canada. Several versions are current. Those with focusing headpieces are particularly useful. Some are sold with a second reserve bulb built into the headpiece. Others have two-filament bulbs, of identical or differing characteristics. Both three-cell and four-cell models are seen. The Edison model P is an older three-cell type using a 4-volt, 1-amp bulb. Its model R-4 is a four-cell model using a 6-volt, 1-amp bulb, although some prefer a BM26G two-filament type rated at 3.7 volts and providing 3 or more candle-power. The Edison model S-1 is much like the R-4, but has an improved headpiece. Both filaments of its bulb are rated at 400 hours. With shorter-lived bulbs, electric cavers often find it handy to tape a spare to the back of the reflector.

All these units are heavy and cumbersome (electric cavers say the same about carbide cavers' canteens). Yet when properly cared for they are effective, and reliable far beyond the limits of "sportsmen's" units. Under ideal conditions they provide eight to twelve hours' continuous light from a full charge. Understandably, this is much less if the 20 percent potassium hydroxide–lithium hydroxide electrolyte is low or has been rewatered with nondistilled water. They are rechargeable almost indefinitely, and their performance is easily tested at home.

For best results, the outside of the battery case should be washed with a bristle brush and a damp rag after each trip, then thoroughly dried at room temperature and stored in a discharged state (some English authorities feel that this is not true of British-made Nife cells). They should be checked for moisture, dirt, and leakage at each recharging, and the electrolyte solution should be changed annually.

The rate of recharging wet cells is critical, and manufacturers' recommendations should be followed as carefully as in other systems. Normally it is about one and a half hours at the same amperage for each hour of use, or one hour at one and a half times the amperage. Most can be left almost indefinitely on a trickle charger at 0.5 amp per hour.

The electrolyte solution used in these cells is a strong, lyelike corrosive. Their seals have been known to deteriorate and leak, causing serious chemical burns and rope damage. After a seal has deteriorated, leakage can occur from mere tilting. Older Edison models are said to be especially vulnerable. Some have suggested wrapping the wet cell unit in several layers of plastic, but Bill Varnedoe says it doesn't help. He recommends replacing the automatic vent with a positive seal, but warns that this causes explosions if not removed during recharging.

When chemical burns occur far underground, treatment is immediate washing with large amounts of water. Some British electric cavers have begun to carry a small plastic bottle of vinegar or other weak neutralizing acid.

Most of these nickel-iron units begin to dim perceptibly about fifteen minutes before they are completely discharged. Canadian Nife cell users report that flickering or sudden death usually indicates a mere loose connection, often corrected by a mere whack on the top of the belt case. The threaded poles on the ends of the headlamp and beneath the fiberglass insulating plate are especially subject to corrosion, and must be checked periodically.

LEAD-ACID UNITS

Although generally considered reliable and powerful, second-hand lead-acid miners' units are reputed to leak more than nickel-iron cells of similar vintage. Comparatively few older models are seen, but brand-new focusable 6-volt M-S-A Minespot ML-1 and ML-2 units (which differ only in their recharging terminals) and Kohler Manufacturing Company Wheat lamps are common in some areas. These are said to be much less susceptible to leakage, and are especially useful in dark limestone and lava tube caves. As in all wet-cell units, distilled water must be added as the electrolyte evaporates (leakage or spillage must be replaced with the electrolyte itself). Bicarbonate of soda and carbonated drinks can be used to neutralize leaking acid.

Current Koehler and Minespot units are sold with a two-filament 4-volt, 1.2-amp bulb with each filament rated at 300 hours. These units can be stored almost indefinitely on a trickle charger at 0.5 amp. They deteriorate irrevocably if allowed to discharge completely for more than a short time.

SPECIAL RIGS: THE VARNEDOE AND OTHER SYSTEMS

Electric cavers tend to be unusually inventive. Going far beyond mere selection for cave temperature, reflectivity, and other obvious parameters, they produce a remarkable variety of personalized home-made rigs.

Bill Varnedoe scorns NiCad D-cells as wasting too much weight and bulk on terminals and case. He also throws away Justrite and most other lightweight battery boxes because of "hopelessly inadequate" battery springs and contacts. When using mercury D-cells in tropical caving, he rejects the standard model and uses the Mallory RM429T because its stud terminals allow him to solder connecting wires firmly in place. Almost always, he chooses batteries with stud and nut terminals rather than what he terms the "abominable" pressure contacts of D-cells. He strongly advocates Gel-Cells and other sealed lead-acid cells with solder tabs.

Other advanced electric cavers use rectangular plastic rechargeable

NiCad cells even though they require special recharging equipment and care. Eveready N76, GE 42B007AA01, and other high-quality 10-amp-hour batteries are often available on the American "surplus" market. Taped together, three of these are about the size of four packs of cigarettes. With a 4-volt, 1-amp bulb they provide eight to ten hours' continuous light. Units of four 5-amp-hour Esse AH4R cells are also common. In long service, such batteries are cheaper than NiCad D-cells, despite their higher initial cost. Both sealed and vented types are available. The former is said to be especially vulnerable to damage through reverse charging of a newly discharged cell by one that is more fully charged. Seemingly identical cells do not recharge equally, and batteries charged in series often contain one or two slightly charged cells. This requires testing of each cell before use and quick action to change cell units when a headlamp dims They are also said to be especially vulnerable to rate and length of overcharging.

Recharging of vented cells is also touchy, even in the upright position. The flat sides must be supported, and vents open. Because their electrolyte level appears low when these cells are discharged, rewatering (with distilled water) should be postponed until several hours after recharging (or so the Diablo Grotto of the National Speleological Society recently reported). Application of Silicon Seal is said to help minimize corrosion.

In the southeastern United States, rechargeable 3.6-volt, 5-amp-hour alkaline batteries have enthusiasts among especially ingenious electric cavers. Their socket contacts are dismaying, but the batteries themselves (Eveready #565) seem to be long lived and trouble free.

Less common but potentially even more useful in large caves with dark walls is the Exide MF2 sealed rechargeable 12-volt lead-acid battery, rated at 6 amp-hours at 77° F. (at the six-hour rate).

LAMP BULBS

Bulb candlepower and bulb life are extremely sensitive to the voltage applied to them. Rarely are they operated at optimum voltage. Merely exchanging carbon dry cells for NiCads, for example, drastically reduces light output. Moreover, so many different bulbs are available that miscalculations are much more likely than in carbide caving. Perhaps worse, watt, ampere-hour, and bulb-life ratings are often far from precise.

Too many bulbs are potentially usable with caving batteries to include them all. This book's section on suggested additional reading (pages 313–322) contains a beginning for those wishing to delve more deeply. The following table represents something of a consensus among the varied conclusions of several manufacturers. But be careful.

NO. OF CELLS	RATED VOLTAGE	CUTOFF VOLTAGE	BULB AMPS	BULB NO.	CANDLE-POWER	BULB LIFE	APPROX. CELL LIFE
				CARBON-ZINC D-CELLS			
2	3V	2.4	.5	PR2	¾	15 hr.	4 hr.
3	4.5V	3.6	.3	PR7	⅞	30 hr.	10 hr.
				13	⅞	15 hr.	10 hr.
			.5	PR3	1½	15 hr.	4 hr.
				365	1½	15 hr.	4 hr.
4	6V	4.8	.15	502	½	100 hr.	24 hr.
			.3	PR17	1	30 hr.	10 hr.
				27	1¼	30 hr.	10 hr.
			.5	PR13	2	15 hr.	4 hr.
				425	2¼	15 hr.	4 hr.
6	9V	7.2	.5	PR12	4½	15 hr.	2½ hr.
3	12V	9.6	.3	993	3½	15 hr.	10 hr.
			.5	965	6	15 hr.	4 hr.
				CARBON-ZINC LANTERN BATTERIES			
1	6V	4.8	.15	502	½	100 hr.	96 hr.
			.3	PR17	1¼	30 hr.	30 hr.
				27	1¼	30 hr.	30 hr.
			.5	PR13	2¼	15 hr.	15 hr.
				425	2½	15 hr.	15 hr.
				RECTANGULAR 10 AMP-HR NICADS			
3	4V	3.7	1.2	BM30	3	100 hr.	8 hr. per charge
			.3	13	1	15 hr.	30 hr. per charge

Bill Varnedoe says I'm a trifle pessimistic on candlepower and too optimistic on bulb life.

In this table, bulbs with a PR prefix have miniature flange bases. All others have miniature screw-in bases. All these figures are necessarily approximate and should be considered only a base line for one's own testing in personal use. Operation at 120 percent of the rated voltage, for example, nearly doubles the candlepower but shortens the bulb life by at least 50 percent and perhaps 94 percent. Fortunately, the rated bulb lives are especially imprecise. Especially if protected against jarring, some technically rated at only one hour often provide several hours' caving.

Self-rescue is possible somewhere around ⅛ candlepower, but at least ¼ CP is necessary to do anything else. Carbide lamps normally emit about 2 CP or a 3- or 4-watt bulb operated as rated, but are not directly comparable to electric units because of their different light pattern and warmer color temperature.

The following list is a handy guide to use of various flange-style bulbs with common hand flashlights. The first type listed provides the brightest light practical, but for a comparatively short time. The last of each group will provide a "survival-type" light for many hours, sometimes even with rapidly weakening batteries. Especially when using carbon-zinc batteries in large flashlights, it is possible to gain many hours of fairly bright light by changing to the next bulb listed in the series below every time the light begins to dim.

2-cell: PR2, PR4 (latter not recommended)
3-cell: PR3, PR7
4-cell: PR13, PR17, PR3, PR2 (latter two not recommended)
5-cell: PR12, PR13, PR3, PR2 (latter not recommended)
6-cell: PR12, PR13, PR3, PR2 (PR18 can be used for short brilliance at full strength)

New quartz-halogen bulb units seem to be working quite well.

Carbide vs. Electric

Some cavers will long continue to argue that carbide lamps throw too yellowish a light and too much sidelight; others, that the electric lights are too cold-looking and throw too much of a beam even with refractive lenses. No matter. Their uses and features differ dramatically, and each is often misused by cavers championing their particular preference. To each his own. The comfortable reassurance of familiarity and dependability overrides much else.

Clearly, we are in the declining days of carbide caving—yet I cannot help recalling one occasion. A girl caver had pooped out—which alone wasn't exactly unusual or alarming, regardless of the tired caver's sex—but this time it was at night, in a nasty chimney in a chill, difficult cave high in the Sierras of central California. Snow was falling in the pitch-black redwood groves. My route splendidly lit by a battered brass Justrite, I scampered to help—perhaps a little too fast and a little too far. A wisp of burning coveralls and suddenly she needed no further assistance.

Every carbide caver has such memories. Indeed, his headlamp is more than a mere light. It boils tea and soup, thaws frozen car locks, welds loose couplings, warms hands (the whole body, if you hunch just right), and performs innumerable other services above and below ground.

Carbide lamps are not the answer for every cave trip, for every caver. But they are among the happiest parts of caving. Today's cavers would do well to prepare for the day when only electric headlamps will survive. But much more than tradition will be lost when the last of us must go electric.

5

The Underground Connoisseur

The unexpected is commonplace in caves. Somebody locks a gate while you're inside, or you lock it yourself and lose the key or find the lock full of dirt when you're ready to leave. Or a stream rises, blocking the entrance passage for a few hours. Or days.

Or you poke your head up a chimney in a streamway cave and find fourteen hours of bone-dry walking passage, and who can resist?

Or somebody gets hurt, and your four-hour fun trip becomes a four-day nightmare.

Even if you're stuck underground for a week, you won't starve to death, but the payoffs of planning can be mighty comforting while you're wondering and waiting and waiting and waiting.

But it's on the ordinary, uneventful trip that being a cave connoisseur pays off the best.

Cave Clothing, Pro and Con

Until 1966 the caving world blandly assumed that clothing was both desirable and appropriate for spelunking. Then an unmistakably illustrated article in *Nude Health Quarterly* boldly challenged our smugness—or at least so it seemed. In sordid fact, however, the intriguing illustrations in that official publication of the American Sunbathing Association were carefully posed in and around Fern Hills Cave, conveniently located in an Indiana nudist camp.

Even though the article thoughtfully included a map of the cave, I haven't had a chance to study it—or its trim trogloxenes—yet. But it would be surprising if its temperature differed from that of others nearby—50° to 55° F., which is really a bit chilly for underground nudism. Clothing is here to stay in almost all caves of the United States and Canada—and further south, too. There, though some, like the Bat Caves of Panama, are warm enough for all clothing to be burdensome, most Latin American authorities have strong views on such matters, and notoriously unpleasant *juzgados!* Moreover, hackly rock, gooey mud, sopping wet or even dry guano, partially decomposed rabbits, pack rat nests lovingly constructed of cactus spines, and innumerable other minor underground nuisances make protective clothing distinctly desirable in all but the easiest "stroll-through" caves.

Except perhaps for photographers' models, comfort and aesthetics are minor considerations. The main role of cave clothing is regulation of the heat exchange between the caver and the cave. Thus, in jagged tropical crawlway caves, the lightest of underwear may be the sole garb beneath lightweight coveralls. On the other hand, high in the Canadian Rockies, in Alaska's Brooks Range, or in the Teton Mountains of western Wyoming, foam-insulated boots and gloves, down parkas, and wool-shielded divers' wet suits may be needed.

The caver's ideal garb—which of course will never exist—should permit instant passage of sweat but not cave water, protect the entire body weightlessly, and immediately adjust to the varied heat transfer situations inherent in the exploration of every cave—even to the kind of ultimate that is encountered in the extensive cave systems in the craters of Mount Rainier. Here, at an altitude that may numb the senses, the caver contends alternately with subfreezing winds and hissing steam vents. This temperature range of more than 100° F. within a few yards verges on the ridiculous, yet emphasizes that all cave clothing is a compromise—for, for those familiar with the principles of heat transfer in caves, even the Summit Caves of Mount Rainier are no overwhelming threat. Instead, they merely require certain special garb superimposed on basic, inexpensive cold cave outfits.

"Inexpensive" is a key word, for clothing doesn't last long in strenuous caving. The tale of Alden Snell, investigating Dead Dog Cave in Virginia during the early days of the National Speleological Society, is only a little unusual.

It seems that the entrance drop was a little tricky, but not bad. Alden led the way, then made himself comfortable at the bottom of the entrance pitch, calling upward. "I don't see any dogs." With that, Bill Stephenson lowered the gear to him, and the others descended, each receiving his pack from Alden along the way. It was not until they reached bottom that they could see Alden sitting on a five-foot pile—a fairly old one, for the cool cave air had no special taint—of dogcatcher's victims. And it was not until they were driving home in a warm car that the dogs made their fate known. Legend has it that Mrs. Snell wouldn't allow him into their apartment until he threw his brand-new coveralls down the incinerator chute, ending the problem.

Or so they thought. Next morning the janitor remarked in passing that some hoodlum had thrown a stink bomb into the furnace. No one seems to remember whether Alden ever confessed the dreadful truth outside caving circles.

HEAT PRODUCTION AND LOSS

The caver's heat production varies enormously. Exploration often demands tremendous bursts of fierce energy—perhaps ten times his resting output—then come long, chill periods of endless waiting in wet, cramped burrows.

These are difficult matters to quantitate. The beginning caver produces and loses more heat than the old hand. Each person, in any case, has his personal basal metabolism rate, and many day-to-day variations affect him as well. Proper clothing cannot substitute for inadequate heat production, nor dissipate excess heat. Nevertheless, it must and usually can mediate between the caver and the cave.

The body loses heat in several ways, which have misleadingly formidable names: radiation, conduction, convection, evaporation, and respiration. Sometimes all methods are badly needed underground. Consider the case of one tired, hot caver during a recent rescue in Arizona's Black Abyss. Panting freely, he exchanged hot breath for fairly cool cave air and lost a little heat by respiration. He unbuttoned his shirt or removed it; his sweat evaporated rapidly in the dry cave air. Although probably unaware of it, he radiated heat to everything around him—rocks, air, unwary fellow cavers. He fanned himself, sought every wisp of breeze, and, maintaining a slow but steady pace, cooled himself a bit more by convection of the air. Passing a flat slab of rock, he flattened himself momentarily against its cool surface and lost a bit more heat by conduction. He daydreamed of dunking himself in cool, clear underground rivers but settled for occasionally wetting his handkerchief and mopping his flushed face, neck, and belly.

Few North American cavers think seriously about heatstroke or heat exhaustion, yet it must often be considered in choosing cave clothing. Heatstroke results from an intolerable increase in the core temperature—simply too much body heat, produced faster than the body can transfer it to the cave. Underground it is usually short lived, yet acute while it lasts. The face and skin flush, the pulse is rapid and pounding. The victim becomes weak, dizzy, and perhaps nauseated. His throbbing headache may be prostrating. The most severe case I have ever seen befell a nationally known caver unwilling to admit he just didn't fit a crucial Kentucky crawlway. Others are reported to have had similar problems Prusiking in heavy jackets donned because they were too cold at the bottom of a pit. Most cavers automatically avoid over-exertion and thus heat exhaustion in tropical or geothermal caves.

In a very few caves, external heat must be considered. Below the cool, muddy entrance passage of Virginia's Warm River Cave, the caver enters what seems to be a delightfully warm passage. Upstream the water and air temperature remain around 92° F., with a humidity of virtually 100 percent. Here the air inhaled is almost as hot as that exhaled, so no heat is lost by respiration, and no evaporation occurs because of the high humidity. The 92° F. cave radiates almost as much heat back at the caver—perhaps more, because its surface area is greater—and conduction is much the same. Airflow is virtually absent; and waterflow, sluggish. Thus the only convection results from movements of the caver, which produce more heat.

Thus the caver leaves behind his heat-producing lamp and moves slowly, mostly in the water, his garb nowadays increasingly limited to a helmet and canvas shoes. Should his pulse race and his face flush, his companions must float him as quickly as possible to the cooler downstream reaches. This is no mere theoretical hazard. Commercial divers overheat when working at 87° F.; at 96° F. if resting.

Cooling the victim of heat exhaustion normally results in rapid recovery to a mere overtired state, but usually ends the day's caving for part or all of the group. Easy venting of all layers of clothing is an obvious need.

PREVENTION OF HEAT LOSS

Conversely, glacier cave exploration requires major conservation of body heat. A slow pace minimizes sweating (evaporation) and convection. Routes are chosen to avoid underground winds almost as much as water, and halts are in alcoves sheltered from both. Heat loss through respiration can be reduced by breathing through the nose and avoiding activity that would cause deep, rapid breathing. Breathing in and out of a wool sock or sleeve helps if conditions are especially bitter. Conduction is minimized by avoiding direct contact with ice, snow, metal ice axes, crampons, and the like. When the explorers sit, they try to do so only on insulated materials like Ensolite. They especially avoid water. Waterproofed down, oiled wool or synthetic fiberpile clothing and full-neck face masks are effective in trapping body heat. But waterproofing is of limited value, and if down gets wet, it can be worse than useless.

HYPOTHERMIA

Hypothermia is a dangerous, potentially fatal chilling of the body's core temperature. Despite recent deaths and near tragedies, this insidiously lethal hazard of temperate and chill caves is still seriously underestimated. Under the euphemistic disguise of "exposure," it has long been the leading killer of unprepared hikers and other outdoorsmen, and many of the needless deaths it has caused have occurred at temperatures encountered in thousands of caves in the United States and Canada. Its prevention depends largely on proper clothing.

THE CORE TEMPERATURE

The body maintains its core temperature through complex processes of heat production and dissipation. In North American caving, it sometimes needs a lot of help. In ice caves, the need for protective clothing is obvious, but in many a pleasant, seemingly safe cave heat absorption may be much greater or much less than the average caver would estimate, and dangerous chilling can and does develop within a few minutes.

Like other members of his tribe, *Homo spelunkissimo* (otherwise known as Joe Caver) can tolerate remarkable chilling of his extremities and skin. Yet he can survive only a small shift in his core temperature—at most some 5° to 7° F. Further, as his core temperature drifts slightly from its normal 99.6° F., his thinking and other abilities begin imperceptibly to deteriorate. He is increasingly oblivious to his own danger signals.

Everyone knows the first phase of chilling, marked by the need to exercise to stay warm. Oral temperatures may already be invalid. Shivering can be controlled but not goose bumps; the first sign that the body is adjusting to preserve the vital core organs at the expense of its surface, and at the cost of a significant energy drain.

At about 95° F. come ominously uncontrollable periods of shivering, an emergency method of heat production. Unfortunately, it is a poor and a temporary defense—weak when the caver is in top condition, fleeting when he is exhausted or injured and really needs it.

If the chilling continues, the core temperature drifts downward more rapidly to around 92° F. At this level the body can no longer maintain temperature defenses. The pulse and blood pressure begin to drift downward, along with the core temperature. The brain loses its reasoning power and the victim is increasingly confused. As his defenses collapse and his core temperature slides toward 85° F., his eyes glaze, and speech becomes slow, slurred, and incoherent. The mind wanders drowsily, forgetfully, and tragically happily. Numb fingers fumble; the feet lose the sense of feel, of balance. Movements are jerky; the victim lurches along and may never move again if he halts. Yet, as long as he slogs onward, he may appear surprisingly normal.

Somewhere past 85° F., the victim becomes irrational. His temperature plummets unless external heat is applied effectively. With shallow breaths and hacking coughs, foamy or bloody sputum rattles in his throat. All too often the pulse slows further, becomes irregular and vanishes permanently, even if healing warmth arrives. At least one

caver has died on his feet, staggering along rigidly, numbly, unknowingly chilled beyond the magnificent adaptability of the human heart.
Others have died in sudden collapse even after their skin and muscle
mass showed an excellent response to skilled rescue techniques. Even
at a far advanced stage, however, hypothermia is usually reversible if
the victim is properly warmed. Below a core temperature of about
85° F., effective action must be rapid, for the victim and his core
temperature hurtle toward unconsciousness, then death, at approximately 78° F.

Resistance to chilling varies considerably. A few have survived
accidental dunking for half an hour in 29° F. seawater without special
clothing. Youth (not infancy) and excellent physical conditioning help
considerably—although they may lead to dangerous overconfidence.

Alcohol is not an antifreeze. It actually increases the rate of heat
loss, and hangovers reduce heat production. Food, hot drinks, hand
warmers, and other artificial heat sources are valuable in the prevention of hypothermia, but proper clothing remains paramount.

THE CRITICAL LAYER

The air, ground, and water temperature of northern, mountain, and
Arctic caverns—and that of the caver's outer clothing—are actually only
indirectly important. Critical is the temperature of the thin layer of air
or water more or less trapped against his body, which is where proper
selection of clothing comes in, for clothing is basically an air trap (or
water trap, in the case of divers equipped with wet suits), and chilling
occurs by loss of this thin layer against the body. Other factors are
obviously important. A caver stuck in a squeezeway will probably lose
heat rapidly by conduction. Yet in much of North America cave clothes
should be planned around the concept of keeping a warm layer against
the body.

WIND CHILL: BEWARE THE BREEZE

Outdoorsmen are increasingly recognizing the danger of wind chill.
That very word, however, lends a false sense of security to cavers,

accustomed as we are to thinking of caves as pleasantly wind free. *At the temperatures of many North American caves, however, wind is not necessary for wind chill.* The minor air currents, which drift unnoticed through most caves, are alone sufficient.

Complex diagrams relate air temperatures and wind velocity to wind chill, but cavers will probably find analysis of a few ordinary situations more useful. Consider wind chill in a 52° F. cave with a gentle breeze, easily overlooked—perhaps around 5 mph. Under these conditions wind chill is roughly the same as that in a cave at 36° F. with barely perceptible wind flow (about 1 mph), or at 20° F. with imperceptible air movement (0.1 mph). This is little problem to a fresh young caver, unless he remains near motionless for a considerable time—or gets wet.

A different situation prevails at places like Jewel Cave's Hurricane Corner. Cooling from an underground wind 30 mph in a cave at 52° F. is equivalent to that at 5 mph in a cave at 38° F. or 1 mph at about 20° F., or somewhere around − 10° F. at 0.1 mph.

Such figures need no belaboring to show that, in temperate or cold caves, cavers' outer clothing should be wind resistant. Fabrics that are totally windproof and waterproof are not suitable, however, for the explorer's sweat must escape. A variety of expensive outdoorsmen's garments that "breathe" are available, but few are suitable for crawlways. Hard-finish, high-thread-count coveralls may be the best available; waterproofing with renewable silicone repellent may be added.

WATER CHILL—BEWARE WATER EVEN MORE!

Despite this hazard, wind chill alone rarely causes significant hypothermia in North American caves. Water chill is a greater threat—for the heat transfer of water (cooling power, if you like) is enormously greater than that of air. Some say up to 240 times as much.

Most find dry, still caves comfortable at 50° F., but 50° F. water is unbearably cold. Without clothing, few humans can long maintain thermal balance in water below 68° F. Even at 76° F., which initially feels pleasant to the swimmer, a naked diver chills dangerously in an

Drilling in waterfall spray for placement of an expansion bolt. The cavers are wearing wet suits beneath coveralls. Note neck protection.

Divers' wet suits worn beneath sturdy coveralls are essential in exploration of many stream caves.

hour or two if forced to remain motionless. Generating maximum heat through extreme effort, an expert commercial diver can tolerate only about thirty minutes without protection at 50° F. Although cavers rarely think in such terms, U.S. Navy divers are required to use thermal protection at and below 70° F. The Thermal Stress Division of the U.S. Naval Hospital at Bethesda, Maryland, reports that some thin persons die after only one hour's exposure at 50° F., and almost every-one after four hours. At 40° F., survival is thirty to ninety minutes. Cavers have been surprisingly lucky—so far.

WET SUITS

For prolonged immersion in cold cave waters, divers' wet suits are virtually essential. Use of three-sixteenth-inch, nylon-lined neoprene garb cures much of the traditional problem of body chafing despite the best that powdering could achieve.

The degree of exposure determines whether a jacket, trousers, or both must be worn. For protection, they are often best worn under coveralls. Wet suit wearers either heat or cool rapidly when out of water, depending on their activity, so wool or other garments recommended below must be added as needed.

Many homemade suits have proven false economy; their seams seem to split at critical moments. Purchase of even the most luxurious models, however, is no answer to the dreadful problem that faces every wet suit user—the inhuman torture of redonning them when cold and clammy.

All wet suits must be washed after every trip—and the key words here are "all" and "every." How such inert material can otherwise become so putrid is one of the great mysteries of caving.

DRY SUITS

Divers' dry suits have not fared well in caves, since they rip so easily. In some special situations, mentioned later, they have limited usefulness.

WICKING AND WATER RETENTION

An insidious form of water chill is commoner and perhaps even more deadly than that of underground lakes and streams: the effect of wet clothing, which holds cold water against the body.

All clothing (unless waterproof) acts as a wick, conducting body heat to the surface, where it is flushed or evaporated in the style of a personal air cooler, albeit an unwelcome one. Each of a series of clothing layers acts as a barrier to this wick or capillary action.

Fabrics differ markedly in their degree of wicking. Wool permits only a little. Cotton and down wick badly, polyester (Dacron, Kodel, etc.) less so, but more than wool. Wool clothing, however, holds much more water by weight than does that of monocellular synthetic filaments. Dacron, nylon, Orlon, and the like thus possess a valuable drip-dry characteristic that can be increased by blotting them with Kleenex.

Until recently, garments of these fibers lacked sufficient pile for real warmth, wet or dry. Now, however, a wide variety of synthetic fiber-pile garments are on sale at competitive prices, and it now appears that such clothing is the best barrier to hypothermia. Some cannot be worn next to the skin, and in this case an underlayer of wool completes the job. In general, wool remains next best, in layers, with wind- and water-resistant overgarments added as needed.

COMBINED CHILL

Beneath waterfalls, high-velocity water and wind team up to drain the caver's body heat at a rate that may be measured in seconds, and swirling spray alongside a waterfall can be nearly as bad. Under these circumstances, heat transfer is so rapid that keeping dry is an overwhelming need. Heavy rubber immersion suits are bulky and cumbersome, but can be lifesaving. Dry suits, rubberized fabric or plastic rain parkas, and trousers beneath coveralls are lesser but valuable protection, since sweating is unlikely under these circumstances. Extra-long lightweight plastic parkas are especially easy to carry and surprisingly durable.

A pioneer Kentucky caver named Floyd Collins faced a different combined chill problem early in 1925, trapped amid dripping water in a slurry of mud and small and large rocks in a tight, blowing crawl-way. Young cavers who may not recall how that event ended may wish to turn to Chapter 10, which considers the rescue problems involved. Similar combined chill is a serious threat in many a less dramatic locale, and is less easily managed than either component alone. A lone English incident permitted some study and, therefore, a certain amount of planning. After three British hikers died of hypothermia due to combined chill, their clothing was tested under a variety of laboratory conditions. Garbed much like many cavers, each had worn a parka, a woolen sweater, a wool-cotton shirt, "string" underwear, cotton jeans, shoes with a single pair of socks, and gloves. When dry, these outfits were found to have about twice the insulating value of a man's summer business suit if the air was still. When sprayed with water in a 9 mph wind in a cool laboratory, nine-tenths of this protection was lost. If waterproof overgarments were added, the wearer was only twice as protected as in dry, still air.

GENERAL CONSIDERATIONS OF CLOTHING

Knowingly or unknowingly, many North American cavers rely more on their exertion for survival than on their clothing. Especially among those who do not often face severe stresses, a motley assemblage of garb is evident. Usually no major problem arises, but a little planning is well worthwhile.

COVERALLS

Preferably waterproofed with a silicone spray and with the hip pockets removed or sewn shut, sturdy coveralls are reasonably standard from Alaska south. Coveralls prevent problems arising from unexpectedly bared beltlines (nobody really enjoys getting caught by the top of the pants when crawling backwards). Arguments between zippered and buttoned styles are past; nowadays zippers almost never inhale so much mud that their nonplussed inhabitants must be torn or cut loose. Gung-ho cavers often reinforce coverall knees by one of numerous techniques: multiple oversewing through layers of innertube, towel, and leather, or pieces of carpet works well. Ideally, such patches should be on a leather base.

Weight and fabric of coveralls vary with the regional temperature. Especially where caves are cold and tight, heavy Dacron types are seen, and snowmobilers' down-quilted suits may be helpful. Because of their cost, however, ordinary heavy-duty cotton coveralls are popular from Alaska to the southwestern deserts. Farther south, thinner fabrics are commoner.

TROUSERS

Unfortunately, cotton jeans are commonly seen in caves. I say *unfortunately*, for they are a latent hazard in other than warm caves of the American Southwest and points south. A bit farther north, wool trousers may make the wearer uncomfortably warm when active. But heat below the waist is normally well tolerated, and opening one's coveralls and shirt usually adjusts the heat transfer. If not extreme, a little such heat early in the trip is good life insurance, rather like donning extra clothing while still warm from exercise.

For cool caves, old wool or drip-dry synthetic heavy-duty ski pants are fine. I personally prefer U.S. Air Force surplus knickers, because they have built-in knee and seat pads. Farther north, down trousers may be vital in extracold glacier caves and in large high-elevation or Arctic ice caves. Merely wearing a second pair of wool pants, however, often suffices.

Plastic overtrousers are cheap, short lived, and fine for waterfall spray. But they don't help when wading.

SHIRTS

Shirts are the least critical portion of cavers' garb. Except in especially cold caves, I usually ignore my better judgment and wear a long-sleeved cotton work shirt, with an unbuttoned wool shirt on top. If the cave temperature is much below 50° F., however, or if water is to be much more than knee deep, wool or the new pile mentioned above is mandatory. Even in warm caves I usually carry a spare wool shirt.

Actually, however, wind chill is so much less hazardous than water chill that getting one's shirt wet calls for rapid reasoning. The nude upper body is easily rid of evaporable water. Drenched cavers may be better off without any shirt for surprisingly long periods; it may even be possible to hurry deliberately all the way to the surface without becoming chilled. Topless caving is risky, but wet shirts may be much riskier.

UNDERWEAR

The new monofilament pile underwear has already been mentioned. Except for this, heavy one-piece wool underwear cannot be surpassed for cold caves or cavers. In the United States, all wool underwear is becoming difficult to purchase. Cavers who are handy with patterns and sewing machines may wish to make their own. In somewhat warmer caves a two-piece set or bottom may suffice, at the risk of a chilly band around the middle. The value of wool and the knit pile is lost if a different material covers any substantial part of the body beneath it. In somewhat warmer caves, however, scant nonwool shorts or what-have-yous are not only feasible but desirable—at least for those who chafe.

Cotton "thermal" and "string" underwear trap some air, and may have a place in long strolls in chilly but dry caves, including particularly spacious lava tube caverns of the Pacific Northwest, but elsewhere I'm skeptical. They wick badly. Beneath wool undergarb this is not necessarily bad, but, caves being what they are, such a combination is rarely needed.

SWEATERS, JACKETS, PARKAS, AND SO ON

Wool shirts or old button-style wool sweaters are an excellent intermediate layer. Coveralls, wool ski pants, wool shirts, and sleeveless wool sweaters permit easy change from heat release to heat storage. Outdoorsmen's water-resistant jackets are sometimes useful atop coveralls, but usually function and survive better outside the cave. Because they trap warm air around the head and neck where heat transfer is comparatively rapid, parkas are more efficient than jackets of similar design. Some modern, hip-length, lightweight plastic parkas are surprisingly durable and of special value in a single water-fall pitch. In very cold caves, mountaineers' down parkas are more useful than down trousers, yet, without any down overclothing, a caver can be comfortable and safe despite ice on the outside of his clothes—providing they are all wool or monofilament pile, and in enough layers.

PROTECTING THE HEAD, HANDS, FEET

In avoiding hypothermia, the most overlooked body areas are the head, neck, and feet. Wet suit headpieces and bootees are cheap insurance, and full-neck wool face masks are available from Arctic outfitters.

The choice of socks varies widely. With boots, two pair of short wool socks are often comfortable in caves too warm for any other woolen garb. The colder the cave, the longer the socks. Indian socks of western Canada are excellent for cold, wet caves, but are expensive and difficult to find in the United States. In warmer, drier caves, almost any comfortable fabric and length is satisfactory.

GLOVES

Gloves are not absolutely essential in caves, but darn close to it. For years I used the cheapest possible cotton work gloves in all but the coldest caves, discarding them every trip or two. Recently I've switched to work gloves with leather palms and cotton backs. These are more durable, more comfortable, and only a little more expensive.

In very cold caves, wool gloves or mittens are obviously important. Since the time my hands were numb in Idaho's Papoose Cave, I've carried a pair in my pack. Mountain Safety Research (MSR) nylon mitts look like a further step ahead, and Ome Daiber is enthusiastic about those made of monofilament pile.

Gloves should *not* be worn when accurate hand perception is vital—in difficult rock climbing, for example, and when using Jumar ascenders (see Chapter 8). The "safety" of a Jumar can be opened unknowingly by a gloved finger.

FOOTGEAR

The majority of serious American cavers wear leather boots, but there is no consensus as to height, style, or sole. Others wear tennis shoes or thick-soled canvas styles. Each has advantages and disadvantages under various circumstances. Canvas shoes (including tennis shoes) are particularly poor on fine clay. They are sometimes lost (at least temporarily) deep in sticky mud. They don't protect the ankles from being twisted or barked, are poor protection against jagged rocks underfoot, and even with wool socks are little help in especially cold water. Yet I prefer them to boots most of the time because of their lightness, low cost, and agility in rock climbing. Thin-soled boat shoes are even better for friction-climbing steep, bare rock, but are much more expensive and less durable.

Presuming that they are waterproofed, the higher the leather boot the greater its protection against icy water, sharp rocks, sprains, and the like. And the clumsier.

Those that lace to a level just above the ankle are especially seen in dry caves, higher ones in stream caves (just why I'm not sure, since even hip waders often fill in the latter). A surprising number of cavers seem content to use ordinary work boots. Few, however, appear to pay much attention to the composition and pattern of the boot sole, despite the special advantages that can be obtained. In traversing slick clay banks, old-fashioned mountaineers' nailed boots far surpass everything else. Yet these have never been common in North American caves because they are otherwise inferior to those with Vibram or other climber-designed soles. I use climber-type boots in preference to thick-soled canvas shoes in exceptionally cold caves, in moderately cold caves with prolonged wading, in long, difficult lava tube caves, and when ascending. Not to mention hiking to the cave through snow, although that's a different sport.

Caves—wet caves, muddy caves, and especially lava tube caverns—are hard on boots. For maximum life they should be carefully cleaned after each trip, dried and cleaned a second time, then waterproofed while warm with a top grade of boot grease. Sno-seal and various new silicone greases have theoretical advantages, but the traditional Hibard's Boot Grease is hard to surpass.

On emerging from soaked boots, a valuable trick is to wipe them inside and out, stuff them with newspaper, and clamp them into a boot tree so that they retain the general shape of the owner's foot while drying. The newspaper packing should be replaced after six to ten hours, and perhaps again on the following day. If needed sooner, it can be done every two or three hours.

PUTTEES

Puttees are especially useful in bat caves, where they keep crawling things on the outside, and in glacier caves, where they keep snow where it belongs. Velcro strips are less effective but better than nothing.

KNEE PADS

Especially in lava tube caverns, cavers often need knee pads badly. Ordinary gardeners' and athletes' pads are not very satisfactory, but felt, leather, carpet, and/or rubber pads sewn onto the knees of coveralls help while they and the coveralls last. "Rockmaster" knee pads are recommended by some members of the Cave Research Foundation. If anyone finds a better style, *please* let my aching patellas know.

Cave Food

Rations obviously depend on a party's plans and the chances of their going awry. Especially in cold caves, or in other caves with a high chill factor, carefully planned food and warm drinks are vital for effective functioning. The variety of increasingly tasty freeze-dried and other compact foods is advancing rapidly. For this book, a few suggestions seem more appropriate than specific menus.

NIBBLE FOOD

From long experience, mountaineers have learned to pack "nibble food" to munch as they hike along. This is probably unnecessary on short cave trips, especially in warm caves and in others requiring little energy output. On the other hand, it's a pleasant habit (liquefied chocolate or tomatoes and similar debacles excepted).

On easy trips, it probably doesn't matter what you nibble: anything that travels well in that particular cave, foods that won't get soggy or unappetizing when dropped or mixed with body heat, water, mud, and crawlways. Foods you can carry in your pocket and get out every few minutes are best. The Personnel Manual of the Cave Research Foundation mentions a tragic mischoice: ". . . bananas whose insides slosh from one end to the other . . ." Hard candy is fine, if it doesn't melt. Lifesavers or hard toffee or lemon drops in a waterproof bag may be better, but they make some people too thirsty. Except in cold caves where the contents may have to be spooned out, I like squeeze bottles of honey wrapped in a plastic bag. These permit a reviving squirt even in the middle of a crawlway, but they do get sticky.

Such foods provide mostly quick carbohydrates and a change of flavor from cave dust. Some find them cloyingly sweet and prefer one of innumerable "squirrel food" recipes often used by mountaineers on longer hikes where more sustained energy is helpful. These tend to be more-or-less edible concoctions largely composed of nuts, raisins, oatmeal and other cereals, and dehydrated fruit, undoubtedly high in protein and fat. The ingredients may be nibbled separately or all at once, or may even be precooked into amorphous lumps that some compare unfavorably with baked guano mixed with stale peanut butter and carbide. The nutritional balance may be stupendous, but forcing the stomach into a rebellious mood is rarely worthwhile underground. Current mountaineers' trends are to palatability rather than nutritional balance except on expeditions lasting weeks or months, or at exceptionally high elevations. Cavers would do well to follow their lead. The healthy body can readily tolerate more severe food imbalances than can be achieved in ordinary weekend cave trips. Personally pleasing packages of raisins or other dried fruits, cheese, cellophane-packed chipped beef, compressed bacon bars, and the like are thus often better than the traditional glop.

CAVE LUNCHES

When combined with the exertion of caving, even a moderate-sized lunch sometimes causes uncomfortable indigestion or muscular weakness. Some cavers nibble all day. Most, however, prefer to pause after a few hours for a more formal meal. Foresight tells here. Consider the tragic piece of cherry pie that descended the Great Pit of Neff Canyon Cave—deepest in the United States—in Rich Woodford's hip pocket.

Good cave lunches are simple, tasty, light, and travel well. They produce only minor garbage, easily repacked for removal from the cave. Some differ from nibble food only because of the greater variety made available by opening the pack. A small can of fruit cocktail or pears can be luxurious—especially if someone remembered to bring a can opener. A juicy tomato in a tin cup—unless a flashlight backed into it. A soft, tasty cheese in a plastic box. Maybe a favorite sandwich in another box, although breads often ride poorly and sometimes attract starving pack rats and other small critters from incredible distances. Canned tuna makes a pleasant cold meal for those who like it, although in some regions it has the theoretical disadvantage of attracting stray felines of undesirable size. Wedge-shaped plastic boxes are made for pie but I don't think I'll ever again recommend cherry pie underground under any circumstances. Beware of bottles, thin plastic or paper bags, and cardboard boxes.

HOT MEALS AND OTHER FOOD FOR LONGER TRIPS

As the length of a cave trip increases, so does the need for fat and protein. Chocolate and nuts are convenient for trips of moderate

length when no hot food is needed. The need for hot meals depends
on the temperature of the cave and the activity of the cavers. In Flint
Ridge—excuse me! It's all Mammoth Cave now—the Cave Research
Foundation has found that in temperatures of 50° to 55° F. one hot
meal is usually necessary on their average eight-hour trip. Ten- to
twelve-hour trips get by with a single meal (usually hot) supple-
mented by an extra can of fruit and two or three candy bars. For
fourteen- to eighteen-hour trips, a second full meal is planned, and a
third if over eighteen hours. These are hot meals in the sense that the
protein source (a can of boned chicken or turkey, Vienna sausage or
the like) is usually heated. The rest of the standard CRF meal consists
of one can of fruit, canned date-nut or chocolate-nut bread, and candy
bars. On some less rigorous trips, cold tuna or sardines are substituted
for the heated protein source. Some western cavers have developed a
special taste for canned (tinned) peas, even when strongly carbide
flavored.

This type of hot meal is inadequate for colder caves and probably
even for comparatively warm caves if the explorer is exposed to special
chill factors. These vary so greatly that it is difficult to specify any
exact interval between thorough warmings by food and drink. At the
onset of any hint of hypothermia all members of the party should
ingest as much hot liquid and food as possible without risking nausea
or excessive sweating. This should be mandatory, for the judgment of
the entire party is likely to be impaired at this point. Sweeten and
flavor to taste only if immediately available, for benumbed minds may
no longer recognize the sugar or Jell-O container. In cold caves a meal
centered around hot soup, hot chocolate, hot tea, hot Koolaid or something of the kind should probably be planned at least every four hours.
During rescues, or at points involving long, chilling waits (belays,
crawlways, and the like), hot food or drink would be welcome for
those stationary and all who pass.

In warmer caves the need for hot foods obviously diminishes, as
does the need for anything more than carbohydrates. With increasing
time underground in cooler climates, with fatigue, and in high elevation caves, fat and protein are increasingly distasteful. Cave Research
Foundation parties often carry canned meats, fowl, or fish for only two

Littering. It was only a small gypsum
cave, but cavers miss it.

meals because few can stomach the stuff after the first eighteen hours
of hard caving.

Prolonged underground expeditions are now so rare in North
America that devoting space here to their care and feeding doesn't
seem worthwhile. The few who plan expeditions long enough to need
a nutritional balance of tasty, lightweight foods can consult standard
mountaineering texts and plan around current dehydrated and freeze-
dried foods. The formidable task of removing a week's trash and
garbage must be planned even more carefully. Burial is not the
answer, and disposal into a single natural receptacle is a very poor
alternative, no matter how convenient and out of the way (see Chapter 9).

UNDERGROUND COOKERY

Palatability is important in the prevention of hypothermia, unimportant in its treatment. When all is going nicely and the caver can relax
and enjoy a culinary halt, the underground gourmet comes into his

own with carefully seasoned recipes. To each his own taste. The choice of group or individual cooking varies with the circumstances. Group meals are more difficult to plan and distribute, but do ensure hot food for potential hypothermia victims.

The heat source is normally much more important than the menu and the cooking utensils. Often a cup and a spoon suffice. Cave cookery can be done with nothing more than a carbide lamp, a pocket knife, and the original tin or can—especially if you use low cans. A candle stub is a bit slower, but several can be used simultaneously. Small butane or Primus stoves—with a supply of fuel in a cave-proof container—are superior and add a considerable safety factor, which often far outweighs their extra weight. Alcohol and other mountaineering stoves are also useful, but many find them trickier to use. Such stoves are invaluable during rescues in very cold caves and in special chill situations requiring large amounts of hot liquids. If pots or pans are used, paper towels are needed to finish cleaning begun by tongue power.

All cave cookery, even heating water, tea, or Jell-O, must be done in well-ventilated areas. Flames and catalytic heaters produce undesirable gases like carbon monoxide (see Chapter 3). Although widely used in Mammoth Cave and elsewhere, Sterno and heat tabs are said to be especially dangerous. Caution is appropriate with all such devices, for at least one underground stove explosion is on record.

GLUTTONY

The spelean gourmet must never allow himself to become the underground glutton, for one man's overindulgence is likely to cripple the entire party at the next crawlway. Light on food, and especially heavy starches, extra light on fats: these are the keys to precaving breakfasts and underground banqueting.

WATER AND OTHER DIETARY NECESSITIES

It doesn't pay to be stingy with drinking water in caves. Even in chilly caves, spelunkers sweat much more than many realize. If drops are running down the face, a caver loses as much as a quart per hour, and dehydration reduces physical performance. A valuable clue is dark, concentrated-looking urine after cave trips. Prevention of dehydration and salt loss is much better than treatment.

In broad areas, subterranean water is immediately available almost everywhere, but almost all such cave water is polluted. Most cavers, therefore, need a pint canteen—often much more than needed, but a valuable precaution.

Cavers underground more than twelve hours will probably have to chlorinate polluted water with halazone or some other disinfectant. Most who have tasted halazone are careful to carry Koolaid or some stronger flavor to disguise its taste. Iodine-containing purifiers are more palatable and perhaps more effective, but too many people develop serious iodism from even small quantities. Don't find out the hard way.

Because of salt losses in sweat, a salt tablet or so should accompany each cupful of water (unless one's nibble food is beef jerky or some other salty delicacy). Wax-impregnated salt tablets are superior to plain. Better yet is gourmet seasoning of meals. Current emphasis on reducing dietary salt doesn't apply in caving.

Connoisseurs' luncheon at the bottom of the Cave of the Winding Stair, California. Since most of the cave consists of vertical drops, even a gasoline lantern was lowered without incident.

Contrary to current speculation, healthy cavers cannot significantly deplete their body potassium even on a prolonged expedition, so don't worry about supplemental potassium or other minerals.

Personal Caving Gear

Water supplies, routine and emergency food and lighting tend to crowd out much that would be handy in the average cave pack. In all but the simplest caves, however, each caver needs a dozen basic items:

1. Helmet with chin strap
2. Headlamp (see Chapter 4)
3. Two other sources of light and waterproof matches
4. Extra supplies for each light source (see Chapter 4)
5. Headlamp kit (see Chapter 4)
6. Compass
7. First-aid kit
8. Pocketknife
9. Reserve food, water, and halazone tablets
10. Canteen
11. Survival kit
12. Pack

HELMETS

Underground, helmets theoretically have several functions besides holding headlamps. Virtually all prevent scalp cuts when the explorer bangs his head. Miners' and construction workers' hardhats are designed to resist short falls and the impact of baseball-sized rocks falling six feet. Miners' types have built-in brackets and are currently the most popular type. If it is necessary to install a bracket on a construction-type helmet, care is essential lest the headlamp point too high. The blade type of bracket, as already mentioned, is overwhelmingly preferred to the tripod. Broad-brim helmets are helpful aboveground, as they give better protection against falling objects, but in crawlways they tend to become unmanageable. A variety of narrow-brim types is available from industrial and mountaineering suppliers.

Supposedly helmets protect the brain when a caver falls, or something falls on him, but few do so, and for this reason more and more cavers are increasingly finding traditional caving helmets unsatisfactory. The ominous sound of even a small rock falling prolongs seconds into interminable hours of sheer terror as those below frantically flatten themselves even farther into almost imaginary recesses. They know that a jagged two-inch rock falling only a few dozen feet will probably penetrate helmet, scalp, skull, and brain alike.

Current impact-resistant climbers' and racers' helmets with crushable liner suspension systems are far from 100 percent protection, for a one-foot rock falling a hundred feet will drive any caver's scalp down to about his navel. Yet they do greatly reduce the commoner risks. Although some are surprisingly comfortable, they are expensive and heavy. Experienced cavers who do little vertical work may not need them. The novice does, and the pit plunger.

Chin straps are a nuisance in crawlways and squeezeways, but essential elsewhere. Recently, a caver fatally hanged himself by his leather chin strap when his helmet wedged in a narrow slot I switched to an elastic type which yields enough for me to tilt my head out of my helmet in such circumstances.

SECONDARY LIGHT SOURCES

Headlamps fail infrequently in caves, but often enough that a secondary light source is mandatory. The Inevitable Law of Inverse Perversity causes the second light also to fail just when it is most needed, hence a need for *two* additional sources. In the past these have been chosen mostly from among candles, flashlights, and hand lanterns. Waterproof matches and cigarette lighters don't qualify, and are considered additional essentials. "Lightning Bugs"—ultraminiature flashlights giving up to forty-five minutes' emergency light—seem promising but data on their shelf life are not yet available.

COOLITE

Coolite's shelf life is two years or more. It is a satisfactory source of emergency light, available as an inexpensive clear plastic tube one-half inch in diameter and six inches long. Bending the tube releases

an activating chemical into a liquid that lights up and fades surprisingly slowly. The bright green light is a bit eerie, but allows one
to see reasonably well for some fifteen feet. Testing by the Central
Indiana Grotto of the National Speleological Society indicates that
it is still useful twenty-four hours later.

CANDLES

The humblest form of cave lighting is the candle. Yet these are a
proper part of nearly every caver's kit. When a party stops for lunch,
far underground, a single candle often supplies all the light needed.
Many a chilled explorer has huddled gratefully over candle warmth,
and many a worried party has returned to daylight by candle power.
Fancy miners' candles emit a trifle more light than ordinary candle
stubs, but not much more. Dripping water and air currents are problems, but some ingenious lightweight, collapsible candle lanterns are
available at outdoor stores.

Like other lights, candles lower everyone's dark adaptation, so their
flame should be kept out of the direct line of everybody's sight.

FLASHLIGHTS

Nearly everybody has a flashlight at home or in the car. Some are
fine to take along into caves. Others are a grave hazard, especially
"glove compartment" flashlights with a cardboard inner liner. Even if
their batteries are okay (which is unlikely), they usually fail as soon as
the cardboard gets wet.

Expensive mine-safety flashlights are the other extreme. These are
designed for use in explosive, methane-thick atmospheres, and cavers
have no need for them unless the cave has been contaminated by oil or
gasoline. This leaves a wide variety of medium-grade models that are
quite suitable for caving. Ideally, the switch should be simple and
sandproof, perhaps waterproof. Plastic bodies tolerate more scuffing
but less impact than those of metal. Bulky "electric torches," "electric
lanterns," or even five-cell flashlights are fine in spacious "walking
caves" (especially in extra-dark lava tube caverns where no light ever

seems adequate). For tight crawlways a two-cell model is usually maximum.

Medium-beam lights are usually best. If your community supports a source for industrial flashlights, such outlets often provide excellent water- and shock-resistant flashlights at a fraction of the cost of prettier but less desirable models. And a spare bulb rated 1 to 1.5 volts below the normal flashlight bulb yields much of the original brilliance after carbon-zinc batteries drift past the 60 percent level.

My own favorite flashlight is an old Eveready industrial model with a demountable switch that can be cleaned and reassembled in total darkness. I use it for picking out features far down passages or pits, for throwing shadows when looking for footholds on slopes and rock piles, and as an emergency light source.

A possible substitute for candles as an emergency light source is the new throwaway flashlight. Their durability in cave emergencies, however, has not yet been tested adequately. And while most hand-pumped flashlights are too fragile for satisfactory emergency use, Bill Varnedoe recommends his heavy-duty all-metal model.

The fail-safe principle for flashlights? At least one extra set of fresh batteries and bulb. Might as well carry them inside another flashlight, just in case you drop the other into a bottomless crack.

LANTERNS

Battery, gasoline, kerosene, butane, and other lanterns are magnificent in throughway corridors of limestone and lava tube caves. Battery lanterns are fine in glacier caves, and maybe flame types if hypothermia threatens—at the expense of ice and its beauty. In stoopways lanterns are a miserable nuisance, and in crawlways, impossible. Even their brilliance is not always of value, for they momentarily blind unwary cavers who chance to look at them. The mantles of gasoline lanterns shatter at disconcerting moments, and the extra fuel always seems to leak into the lunch. I've owned one for twenty years and never used it in a cave. Sometimes they serve well as one of the extra light sources—but not often.

WATERPROOF MATCHES

Vital to every caver's pack are waterproofed matches or ordinary wooden or wax matches in a waterproof container. Mountaineering suppliers offer a considerable variety. A reliable striking source is equally essential. Whatever the chosen combination, it must function when everything (including the caver's hands) is sopping and muddy. This is another area where advance testing is highly important: a cave is no place to learn that one's waterproofed matches are also fireproof. My own preference is a widely available plastic waterproof matchbox kept about half-full of ordinary kitchen matches, plus a piece of striker surface loosely folded *away* from the matches. I also carry an ordinary book of paper matches (wrapped in a plastic bag) in a convenient pocket, and rarely have to dig out the others. Butane cigarette lighters of the type with visible fuel—if you have one that works every time, in rain and mud—can also be used.

COMPASSES

Modern outdoor stores stock a bewildering array of compasses. In caves two basic uses determine the selection: (1) finding your way around and out, and (2) mapping, which is an integral part of cave science and hence outside the scope of this book. Brunton and other precision surveying compasses produce exceptional cave maps, but are less than ideal for finding one's way out of a complex. I personally prefer a simple "lensatic" compass, but many others that function easily in tight holes are equally satisfactory. They are especially useful in labyrinths and in long, straight passages where it is easy to get turned around 180 degrees. Both foresights and backsights should be taken, for a clear-cut fail-safe principle applies to compasses in caves: Don't trust them very much. This pertains especially to lava tube caverns but also to some limestone and other caves, where magnetic materials in the bedrock or one's own equipment can produce disconcerting errors.

On the other hand, if you're tired, cold, and hungry and the compass disagrees with you about the way out, don't automatically assume the compass is wrong. Nobody may ever see you again.

FIRST-AID KIT

At one time I carried an elaborate but compact army surplus first-aid kit that included surgical instruments and a plasma substitute. After it was stolen at the entrance of a West Virginia cave by some member of a skylarking local Sunday School class, I rethought the situation. Now my kit includes six Band-Aids, six butterfly tapes, a roll of one-inch adhesive tape, a small package of Kleenex, a clean hand-kerchief, a three-inch elastic bandage, a rolled wire splint, a small bar of soap, a needle, and six aspirin tablets (nine to eighteen hours' supply).

This is more than a little heretical, but there is method behind it. Underground, a clean, ironed handkerchief is almost always as good and sometimes better than sterile dressings. Torn clothing can be used quite well for outer bandages and tourniquets. Soap is at least as good as commercial disinfectants, and less likely to be misused on open wounds. Morphine syrettes are too likely to be stolen, and Joe Caver can't carry them legally anyhow.

The rest is pretty obvious. Snake-bite kits and more elaborate first-aid kits are needed for special situations, discussed in Chapter 9.

POCKETKNIFE

Being basically a surgical instrument, this perhaps should be listed as part of the first-aid kit. Yet a knife has many other uses under-ground, especially as a cooking and eating utensil. Any clean, sharp, folding type will suffice, although those with an additional can-opener blade are extra useful. Some choose the type with a dozen folding miniature tools. Because they come out of their sheaths too often, sheath knives are not an acceptable substitute (see the section on vertical caving, page 230).

Rather obviously, the blade should be cleaned after each use. Then it can be used to remove fragments of wood, and so on, after merely flaming it well with a carbide headlamp.

SURVIVAL GEAR AND BIVOUACS

Small, compact survival kits are available from many mountain rescue organizations and outdoors outfitters. Most are designed for mountaineering, however, and their contents are somewhat superfluous for caving. Perhaps their most useful component is the body-sized survival tube. Constructed of lightweight plastic, this is designed to conserve body heat in a bivouac. Two garbage-can liners serve almost as well.

Bivouacking cavers occasionally have slept comfortably and happily far underground. Unfortunately, this rarely happens except on large expeditions outside the scope of this book. At least in cool to cold caves, the average cave bivouac is a debilitating struggle against hypothermia and sleeplessness. Bunched together to conserve each other's body heat, each bivouacker half dozes in a painful crouch designed to absorb every possible calorie from a heat tab or carbide lamp.

Sleeping bags and ground cloths can be an enormous comfort, but their burden can be disastrous. See my *Adventure Is Underground* for the unhappy details of a nightmarish thirty-three-hour struggle caused by our efforts to fit them through the upper squeezeways of Neff Canyon Cave. Down bags should not be used if there is any chance that they will get wet. After wringing out excess water, fluffing of bags filled with new synthetic fibers like Polarguard, or Thinsulate, or Hollofill II restores almost 100 percent of their original insulating ability.

If possible, all wet, wicking garments should be replaced or removed. Wet boots are a particular problem, and the feet may do better inside a rucksack. Pockets should be emptied. Clothing is ideally adjusted to be loose around the body but snug (not tight) at the wrists, ankles, and neck. The neck and head often need special protection against heat loss. A knitted skier's wool face mask is one solution,

Plastic windbreak in the drafty corridor of Overholt
Blowing Cave, West Virginia.

and breathing through a wool sock or sleeve helps conserve body heat.
So does an Ensolite or other foam pad to reduce ground conducting.
One should also select a drip-free, rock-free bivouac area. Some are
enthusiastic about "space blankets"—aluminized polyethylene insulating film laminated to both sides of a fiber core.

Even in grim surroundings, foresight leads to something vaguely
resembling comfort. The Ensolite pad helps as does hot food or drink;
a plastic windbreak; a few plastic bags, one stuffed with Kleenex to
blot up water; skiers' pocket-sized handwarmers (Ome Daiber recommends the type using solid fuel); perhaps even a pocket-sized bottomless plastic shelter that traps the body heat of the entire party, using
cavers as corner tent poles. Such small items can make an enormous
difference.

OTHER CAVING EQUIPMENT

Extra bootlaces or shoelaces have many ingenious uses besides their
normal function. Paper clips and safety pins are surprisingly helpful,
too: I carry the latter on my helmet, with a tip cleaner hooked to it.

Pencils and notepads may be difficult to manage. I carry a couple of stub pencils in convenient coverall pockets, and sometimes my notepad in a plastic bag in my helmet.

CAVING PACKS

The essentials, lunch, extra clothing, and a few comforts can be compacted into a pack about the size of a football. Besides ordinary small packs and rucksacks, heavy-duty bags available at surplus outlets serve well. Some prefer to make their own from gallon-sized plastic jugs.

Most cavers prefer styles worn on the back, but some choose shoulder-loop or belt styles easily removed for dragging through crawlways. In general, waterproofing individual items inside a pack is preferable to attempting to waterproof a pack. Far underground, few sights are sadder than a water-filled waterproof pack.

Obviously, no caver's pack is adequate if some essential is forgotten, and the list should be checked before each trip. The life of an entire party may depend on it.

Underground Conviviality

Vodka burns well in carbide lamps and alcohol stoves for a considerable time but is said to gum up the works eventually. All other alcoholic beverages cause worse lamp problems.

Yet this is a better underground use for alcohol than its convivial one. A single short beer probably never hurt anyone, but at least one recent American cave death was due to beer. A single bottle may cause some unfortunates to lose the fine judgment often vital in caving.

Theoretically, a modicum of spirits would serve as a happy tranquilizer at bedtime on a knobby dirt floor. Unfortunately, it also increases urine flow, leading to dehydration, salt loss, and contamination of the cave. Too, the temptation of having more than a modicum is inevitable.

Nor is such concern pure theory. Bill Cuddington reports that one southeastern caver with a different philosophy has twice had to be restrained from rappelling with a carabiner blearily snapped around his belt instead of into his Swiss seat. Come to think of it, that noted connoisseur Bacchus never figured in the Graeco-Roman myths of the underworld.

The Best Is None Too Good

Being an underground explorer is making the best of the worst. Eating what is available and appealing when bone tired, muddy, and sweaty, all at the same time. Adding and retaining just enough heat. Using stoves where safe. Carrying the essentials plus a few scant comforts and drinking enough water to avoid dehydration and salt loss. But it's a lot worse if you're not a cave connoisseur.

Mammoth Cave is of special historical and geological interest, as well as qualifying as the world's longest cave (about 150 miles mapped to date). This "throughway" passage displays well-preserved wooden pipes and vats used for leaching saltpeter from cave earth around 1812. Exposed in the domed ceiling are bedding plane anastomoses, important phreatic speleogens.

6

Cave Ropes and Belaying

Feet first, shirt over head and more than a little skinned and bruised, young Israel Putnam, the young hero-to-be of the American Revolution, hurtled howling from the little Connecticut cave. The first recorded North American cave belay had worked a little too well.

As today's caver knows it, belaying is the art of rope handling by one person to prevent or minimize the effects of an accident that might befall another. Usually this is to halt falls by friction of the rope around the belayer's body. Yet odd situations somewhat like that of Israel Putnam still occur. His was at least a sort of belay—a rope tied around his middle as he crept, birchbark torch in hand, to see if a livestock killer was at bay in what is now Wolf's Den Cave. Few details have survived, other than that his friends overdid it. At the gleam of desperate eyes in the blackness, Putnam tugged on the rope as arranged, and was hauled, seemingly, halfway into Massachusetts. It worked even better the second time, for as a youth Putnam was no less outspoken than as a famous general. Much like today's cavers, he instructed his belayers vigorously, sulfurously, explicitly—and successfully—before he crept back in to shoot the cornered killer.

For more than two hundred years after Putnam's famous belay, the use of ropes in caves changed surprisingly little. Today, however, spelunkers use them much more broadly: for belaying, for rappelling, for standing rope ascents, and in a few other ways to be mentioned later. Especially in the past two decades, newly evolved techniques have revolutionized caving, creating whole new concepts of exploration.

Yet we still relate to the hero-to-be of the American Revolution and his 1743 belay problem.

Cavers' Standing and Belay Ropes

In the 1940s manila rope was the best that American cavers and mountaineers could find. So we all used and misused three-strand ("laid") manila rope, mostly $\frac{7}{16}$ inch in diameter. Nylon rope of similar construction came into use in the 1950s. Besides being stronger and less subject to deterioration, nylon was pleasanter to handle. Rappelling on its springiness was delightful. Yet with time other problems came to light: for example, nausea from spinning in midair. And if a belayee slipped when ten feet from the bottom of a 150-foot descent, the stretchy nylon lengthened ten feet, hardly slowing his fall at all.

At this point, mountaineers' and cavers' rope needs began to diverge. Both appreciated the nonspin features that came with newer synthetic "kernmantel" ropes, with a woven sheath and an inner core that carries the load. But mountaineers need stretchiness for "belaying the leader" (see below), as cavers rarely do.

For most caving uses, an extremely low-stretch nylon rope was needed. Vertical cavers were unable to persuade rope manufacturers to produce such a rope, so one of them—Richard Newell, a caving textile engineer—personally designed and began to manufacture Bluewater II. Quickly it became the world's best-selling caving rope. Others followed: Bluewater III and some successors, Pigeon Mountain (PMI) rope, and some European varieties now beginning to be sold in North America. Bluewater III is slicker and a little harder to learn to use. Like the popular new PMI rope, it probably is a little more abrasion resistant than Bluewater II. For caving all are vastly superior to the ropes of the pre-Bluewater days, and to a considerable degree, today's vertical cavers select their rope to match their own needs and personalities.

For sling ropes, discussed below, other synthetics are often preferable. For the standing rope, stretch free ropes are in a class by themselves.

Whatever textile is chosen for belay and standing ropes, they must be thick enough to be strong, durable, dependable, and reasonably

comfortable to the caver on each end. Ease of transportation is a lesser consideration. Ropes of 7/16-inch diameter are commonest in the United States and Canada. Especially if the belayer is well padded and several coils encircle the belayee, this size causes comparably few burns and bruises. Rope of ½-inch diameter is even better but considerably heavier and more expensive, while ⅜-inch rope can be used where miles of hiking require ruthless elimination of every unessential ounce. Its tensile strength can be adequate, but it cuts into the flesh. While this can be controlled by tying into a body harness instead of body loops, cavers trying it in belay practice rarely use it again.

CARE OF ROPES

The chief enemies of rope life are sunlight, dirt and other abrasion, moisture (not much of a problem with synthetic ropes), battery acid, vinegar, ketchup, salad dressings, bleaches, household cleaners, and a surprising number of other common chemicals. They are also subject to damage due to overheating by rappelling devices, furnaces, and jammed washing machines, among other sources. Some of these threats are self-evident. Others are discussed in the next two chapters.

Kyle Isenhart prepared and published a well-researched report on safety factors in the care of nylon caving ropes.* Among his recommendations are:

1. Using ropes only when dry whenever possible, even for rappelling
2. Avoidance of all heat above 180° F.
3. Avoidance of sunlight, especially in hot places like automobile back window shelves
4. Pre-use washing and drying
5. Frequent laundering to the point that the rope appears clean

* Published in *Georgia Underground*, vol. 10, no. 3, May/June 1973, pp. 87–93.

Ice forms a variety of speleothems. Shown here are frost crystals, dripstone, and flowstone.

6. Use of commercial front-loading washers with glass doors
7. Drip drying in the air (some commercial dryer drums get too hot)
8. Washing in a loose bundle rather than in a coil (despite resulting snarls)
9. Special care that all the rope is inside the washer drum
10. Use of detergents rather than natural soap (liquids: Wisk, Tide and similar types; powder: Tide)
11. Avoidance of chlorine-type bleaches
12. Addition of nylon fabric softeners (final cycle rinse types: Downy; initial additive types: Johnson's Rain Barrel)
13. Avoidance of formic and sulfuric (battery) acids and phenol (carbolic acid); the latter found in various wood preservatives

Veteran vertical cavers can often be spotted by the way they treat their ropes. Legend has it that "Vertical Bill" Cuddington never allows a rope to touch the ground. Actually, he does sometimes place a coil on the ground, but he is careful that his ropes never drag, much less get underfoot.

When washing ropes, soap or detergents appear to be generally unnecessary. Rinses should be repeated until no further dirt appears in the rinse water. A water temperature of about 110° F.—"hand temperature" —is widely considered adequate; indeed, if hotter water is used with "unprocessed" controllable stiffness rope, it comes out "processed" and stretchy. George Martin, Jr., recently had an unusual problem with a front-loading commercial washer-dryer that had a plastic window. His 600-foot Bluewater II somehow stuck to this window. The spin-dry cycle melted a wide strip of the sheath and left him with one 400-foot and one 199-foot 10-inch rope. This might not have happened with a shorter rope, a glass door, or a center-post washer. Nevertheless, his suggestions of caution, vigilance, and padding the window with canvas are well taken.

WHEN TO RETIRE A ROPE

Dynamometer testing may not be necessary for ropes whose owners know every detail of their history and inspect them carefully before

Some limestone caves are found in alpine deposits of grainy marble.

each use. Mountain Safety Research recommends retiring kernmantel ropes when a ten-power magnifying glass shows that 60 percent of the sheath strands are disrupted where the rope looks most worn. If necessary, the fuzz should be melted back for better inspection. Otherwise it should be left intact, since it protects the rope.

Organizational ropes often pass through the hands of cavers who may be careless about exposing them to gasoline fumes, abrasion, and other hazards. Such ropes should undergo careful inspection under 1,000-pound tension (measured with a dynamometer) before each use. This is easy with tensionless anchoring (see Chapter 7), a car and a tree.

PREVENTION OF RAVELING

Unlaying or raveling of kernmantel ropes is much less of a problem than with laid ropes, but still merits attention. Palm-and-needle whipping with small cord can be used on both. Some cavers merely use friction tape wound tightly around the rope, but flaming is a better way to prevent separation of the sheath from the core of the rope. The end of the rope is heated until slightly pliable, then twisted with a leather glove, forming a tapered cone with a length of two or three times the diameter of the rope. If the rope is a laid type, the twist should tighten the lay. A section of rope before the cone should be lightly flame-sealed and should extend three times the length of the cone. This section should not be heated to the point of stiffening the rope.

Slings

In some ascent systems (see Chapter 8), not all types of sling ropes work equally well with standing ropes of different textiles and weave, but with Bluewater and PMI standing ropes, braided Tenstron or other polypropylene sling ropes have become standard. In general, slippery-feeling sling ropes seem to be best for slippery-feeling standing ropes, and braided sling rope is usually superior to laid. Nonstretch slings are theoretically desirable, but, for sling on a Bluewater II standing rope, solid ⅜-inch braided polypropylene functions especially well.

For most semimechanical knots, nylon seems to be especially good. Ordinary polypropylene sling rope is a bit stiff for the helical ascent knot, but a special laid form named Tenstron works well. In some other knots, on the other hand, Tenstron jams comparatively easily. Although said to be brittle where bent sharply, it is considered especially good for foot slings. For the helical knot, some types of kern-mantel nylon sling rope work well in dry conditions. These have the advantage of being stronger than "poly" ropes.

Some alternatives are confusing, and may cause dangerous errors in difficult conditions. Ropes of ⁵⁄₁₆-inch polypropylene and perhaps polyethylene slings are probably strong enough for foot loops but not for chest loops, for example. Hollow polypropylene rope is sometimes seen, but many feel that it should be limited to cave diving because of its excessive stretch and slippage—although cave divers may also demur.

Because of its strength, ease of handling, and low cost, flat webbing is widely used for foot and other slings. Some ascent rigs use ⁵⁄₁₆-inch Bluewater II rope for a long foot sling running through an ascender box (see Chapter 8), and doubled ¹⁄₃₂-inch webbing one inch wide for a short foot sling. Single webbing of this type has broken under cave stresses, and, even when doubled, webbing must undergo adequate tying and stitching, discussed below. As a result of recent accidents due to its failure, yellow tubular nylon webbing is increasingly shunned.

Whether rope or webbing, slings should not pass directly through any narrow or rough metal opening. Rope thimbles can be used if no better alternative exists.

Body Harnesses and Seats

Most advanced vertical caving techniques require some form of special harness or seat. This is even being applied increasingly to ladder climbing, as indicated in the next two chapters.

Simplest is a loop of ⅜-inch sling rope about six feet long, twisted once to form a figure eight. Such a loop can be set under the gluteus maximus to form a rudimentary seat, or used in the manner of a Swiss seat (described below). It can also be passed around the shoulders to form a shoulder harness held together by a carabiner. When stressed, however, such rope harnesses and seats soon become uncomfortable. Some cavers have purchased climbing harnesses and modified them to their special needs. More use webbing of various widths, creating innumerable styles of harness and seats.

Swiss seats, which are comparatively standard, are close-fitting loops of tied or sewn 1¾- or 2-inch webbing that form a sort of open diaper in which the caver sits. To form a Swiss seat, the caver snaps a reliable locking carabiner onto a prefitted loop and allows it to dangle behind him. He brings the carabiner and the bottom of the loop forward between his legs, then pulls the sides of the loop around each side and snaps them into the carabiner, too. Automobile seat belt material and soft-weave nylon parachute webbing are especially comfortable.

A simple chest harness is easily formed from a wide piece of webbing about eight feet long. One end is held at midchest, and the other is passed around the chest under the armpits, then on around, up and over the shoulder, and back to the center of the chest, where it is securely buckled. A carabiner is snapped around both loops, and another may connect it to a Swiss seat carabiner. Because of recent carabiner failures, more and more cavers are using two in each position.

If adequate D-rings and other hardware are used the stitching is the weakest part of the harness or seat. Use of cotton thread and inadequate homemade stitching have caused several serious harness failures. Stitching at rubbed points tends to go bad; the thread loops may be worn off and the harness then pulls apart easily. Heavy-duty nylon sewing machine thread should be used, and the power sewing machines found in shoe repair shops are strongly recommended. John Cole insists on 480 stitches (sixty inches of sewing) of untwisted type D nylon thread, while I, being more than moderately chicken, use heavier type F

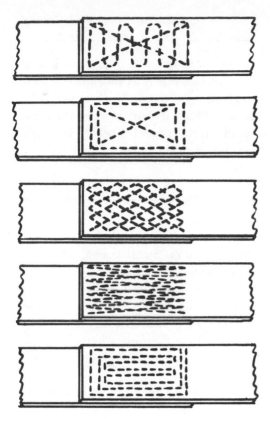

Fig. 1. Several styles of stitching for harnesses and webbing.

thread for my Swiss seat. As far as the stitching itself goes, several patterns work well. A long back-and-forth rectangle with crisscross diagonals is especially popular. Some prefer box-X and double box, double W patterns. A few add "wear pads," lightly stitched atop the oversewn areas.

Belay Techniques

Unlike mountaineering, most cave belaying is static. It requires keeping the rope just barely taut, thus halting a fall before the victim gains significant momentum. While a skilled belayer makes the technique look deceptively simple, surface belaying practice is essential before any underground attempts. The first time a learner tries to stop someone dropping only a few inches with a static belay, he learns much more than he expects. If the practice fall is as much as a foot— or, far worse, a yard—his learning may be painfully fast. But he'll never again let anyone fall that far. Which is just as well. On non-stretch ropes, a statically belayed fall of fifteen feet is likely to kill both belayer and belayee—no matter how excellent the belay.

BELAYING THE DESCENT

Nearly all serious cavers are thoroughly familiar with belaying, and teach its fine points to each new generation. Small points, however, sometimes get lost, so frequent review is quite appropriate.

Every caver must be able to belay equally well right- and left-handed. For purposes of illustration, let us choose the left, with the untied coil of rope placed on the belayer's left in such a way that it will run free without snarling. The belayer passes the rope around his back to the caver to be belayed. (Details of tying in and positioning are covered below.)

As the belayed caver begins his descent, the rope slides through the belayer's left hand, which is held *close to his navel,* thence around his left side and his back and onward to his fellow. At all times the belayer is alert. Should the belayed caver stumble as he approaches the descent, the belayer's left hand is ready to whip the rope far across his lower chest and into the right armpit. The braking hand (in this case the left) itself contributes little to the friction that halts the fall. Its role is to maximize the body friction by positioning the rope against as much body surface as possible, as tightly as possible.

The brake hand also controls the speed with which the descending caver can take out the rope. Unless the belayee specifically requests slack in the rope (dangerous but sometimes momentarily necessary), the belayer keeps a slight tension on the rope at all times. This allows him to feel movements of the caver below, and also reassures the man below that the belayer is alert. The belayer rarely has to worry about pulling too hard, for the climber will inform him immediately and forcefully if tension on the belay rope is bothering him.

Occasionally climbers need extra tension on the rope. Often this can be provided by the belayer merely leaning backward, or moving his brake hand into brake position. Sometimes the other hand must be used in lock position, as described below.

In belaying a descent, the nonbrake hand is best used as a brace, stiff-armed. Otherwise it rests lightly on the rope somewhere around the level of the hipbone. If the brake hand has gotten out of position (dangerously back or outward from the body), or if slack must be taken in, the nonbrake hand can be used to lock together the portions of the rope in front of the belayer. The brake hand then adjusts the rope. Closing the loop in front of the belayer's body in this way cannot overcome the momentum of a falling caver but can hold him in place, dangling in midair, for a considerable time. The brake hand can be slid along the rope to rid it of dangerous slack, and again be prepared to brake at any odd moment.

When slack must be taken in, the lock hand feeds it backward, around the belayer's back, to the brake hand, which in turn feeds it to the rope coil. In doing so, the brake hand must never pass so far away from the body that the lock hand cannot lock the loop beyond it. When considerable slack must be taken in rapidly, short, fast feeds by

Some celebrated caves have long been developed as tourist attractions.

the lock hand are essential. Care in the original position of the rope coil is as important as careful placement of the slack added to the coil, because a snarled rope in mid-belay can be real trouble.

The main role of the belayer during a descent is controlling the rope taken from the coil by the caver on the end of the rope. Belaying during an ascent is quite different. The main task is taking in slack as the climber advances. Sometimes he climbs so rapidly that a large, loose loop forms behind the belayer's back. This is particularly dangerous. If such a loop dangled, the climber and others could see it and he would halt, yelling "Up rope!" Here, however, it is a hidden menace. The belayee continues to climb, unaware that if he falls he will gain considerable momentum before the belayer can do anything. If the rope cannot be kept taut against the belayer's back, he must yell for the climber to halt while he adjusts it. Unfortunately, the climber cannot always comply safely.

Occasionally a caver may become too tired to continue the ascent. Voluntarily, or willy-nilly, he may put part or all of his weight on the belay rope. The time he can hang is variable. If the pull is distributed by some form of body harness or seat and other factors are favorable, he may be able to dangle safely for many hours—longer than the belayer can tolerate. If he is hanging by a single chest coil, or even two or three, his safety may be measured in minutes. A well-placed belayer should be able to support all the climber's weight for a short time without excessive discomfort. All concerned will fare better, however, if the belayer demands that the climber proceed within two or three minutes.

Even if a belayer never faces the need to hold a friend in midair for hours, his comfort is extremely important. Most cavers pass the belay rope around especially well-padded portions of their anatomy after just one good rope burn. Cloth gloves with leather palms also become popular very rapidly.

BELAY POSTURE AND ANCHORS

The belay position is at least as important as rope handling, for the belayer must not be pulled off. To find oneself jerked upright from a seemingly secure position is terrifyingly instructive. For endless milliseconds you stare straight down into nothingness, still stiff-legged, ludicrously still bent in the middle, sitting precariously in midair, wondering helplessly if you're going to tip on over. When it happened to me, I was lucky (not to mention the guy I was belaying). I slowly sank back into the original position—which suddenly didn't seem so secure. But I learned: Unless the position is so secure that the shock of a fall will only wedge the belayer still tighter, *all belayers should be anchored or themselves belayed.*

The posture for each belay is that which will best take the shock of a fall in that particular place. Almost always this is in a sitting position, facing the climber. The legs are braced stiffly and securely under each side of the rope. The feet should be as high as possible without risking the rope's sliding under the body. The locking hand and occasionally a shoulder and even the head provide a third bracing point. If the rope must pass to one side of the feet, the third bracing point must be across the rope from the feet. The ultimate ideal converts the belayer into a sort of animated crossway log, directly over the fall line of the belayed caver and braced at both ends.

Anchoring is discussed at some length in Chapters 7 and 8. Suffice it to say here that for belaying anchoring is especially simple and quick if the belayer wears a Swiss seat backward. Into his seat carabiner is snapped a rope loop anchored to whatever is convenient—a tree, a rock, a piton, a well-wedged fellow caver. If the belay position is even moderately good, most of the shock will be absorbed by the belayer's body, and the rope loop can be ordinary ⅜-inch sling or webbing. If the position is exceptionally exposed and the anchor will take most of the shock of a fall, ⁷⁄₁₆-inch rope should be used.

Obviously, the belay position must be chosen to avoid running the rope over sharp rocks. Padding should minimize danger of abrasion, and placed so that it cannot be dislodged by a fall. A good friend of mine died needlessly from a minor fall in the mountains, his rope neatly cut when his belay padding somehow slipped off a sharp ledge.

THE STITCHT BELAY PLATE AND LINK

In some difficult situations an inexpensive new device offers considerable assistance—the Stitcht belay plate or link. The original model is a simple metal plate with a hole that permits insertion of a short loop of $\frac{7}{16}$-inch rope, to which the climber ties in the usual way. The loop is passed through a carabiner, which is attached either to the belayer's anchor or to his waist. Feeding and taking in rope is easier than in a body belay. When a fall occurs, the belayer hauls back on the coil end of the rope. The plate or link snaps close to the carabiner and friction takes over. Hand control is easy. This device is said to function better with braided or woven ropes than with hard-lay types like Goldline.

BELAYING FROM BELOW

To belay from below, twice as long a belay rope is necessary as a pulley is anchored at the top of the pitch. The last man to descend feeds the short end of the rope through the pulley and ties on normally. He drops the long end to his belayer below, and descends just as he would with a belay from above. Carabiners, descending rings, and similar devices are poor substitutes for a pulley.

When belaying from below, anchoring the belayer may be even more important than when belaying from above. Consider the yarn of four southwestern cavers who discovered a tight crack that "went"— into the top of a large, beautiful cave seventy feet below. After rigging a pulley belay and a cable ladder, they rappelled happily and explored the cave for several hours. The first man back up the ladder, however, was heavy, and his belayer slight. Just as the climber reached the top of the ladder, it broke. The unanchored belayer was rudely yanked into the air but hung on manfully, even when the climber collided with him in midair, breaking his left clavicle. Only when stunned by the ceiling (despite his hardhat) did he release the belay rope, which, of course, whipped through the pulley. At the same moment, the ladder climber broke an ankle—on hitting the floor. The belayer lit on him and the broken ladder. Fortunately the snarl cushioned the impact enough so that he broke only eight ribs, two arms, and the climber's pelvis. This really could happen—one reason more and more cavers often use a self-belay on a standing rope (see Chapter 8).

BELAYING THE LEADER

The problems of belaying the first caver across the edge of a pit, or up a rock wall, however, are far from legend. No matter how good an ordinary static belay, such a caver will either pendulum, smashing into the rock below the belayer, or fall past him, quickly accelerating to a fatal velocity. In such situations the direction and length of the potential fall must be changed. In mountaineering, this is done by running the rope through a carabiner attached to a piton, chock, or expansion bolt hanger (see Chapter 7). The belayer also allows a little of the rope to run while stopping the fall—a dynamic belay. When all goes well, the falling caver comes to a halt over a period of a second or so, ideally before he hits anything excessively solid. In caves, the process can sometimes be speeded by using solidly anchored stalagmites and smooth projections instead of pitons or bolts. The rope can be passed around such anchors, or through carabiners anchored to them by sling ropes. Unlike mountaineering dynamic belaying, this is a tricky, dangerous technique, best left to highly skilled experts who have taken much more than their share on both ends of practice falls. Those getting into such advanced matters may find useful the Mountain Safety Research auto-belay device and the carabiner friction hitch (Italian friction or Munter hitch) described in mountaineering reports.

WHEN TO BELAY

Although of enormous value, belays are time-consuming and thus somewhat hazardous. They are used in and around caves whenever they significantly reduce the chance of injury or death—not just injury or death from dangerous pits or rockfall, from errors in judgment or technique or equipment failure, but from faintness and fatigue, or anything else. They should be used on every ladder pitch, for example, since today's cave ladders are never so long that the weight of the belay rope would be excessive. The value of each must be balanced against the need, with the decision always on the side of maximum safety.

Sometimes a novice eyes an easy pitch uncertainly, then requests a needless belay. If refused or belittled, the abashed beginner usually conforms and all goes well. Next time, however, he is likely to fear further criticism or ridicule, go beyond his ability, then panic and fall needlessly. Every request for a belay must be treated seriously. If a better alternative exists, the person requesting the belay must be made so comfortable in accepting it that he will have no hesitancy next time he wonders about asking.

BELAYING RAPPELS AND STANDING ROPE ASCENTS

In the early days of standing rope ascents, near-disastrous rope snarls led to the belief that no such ascent or rappel should be belayed. The development of nonspin ropes has altered this drastically. If both the belay rope and the standing rope are Bluewater II or its nonspin cousins, the belaying of ascents is limited chiefly by the weight of the rope—around two hundred feet.

With such ropes, rappels of about two hundred feet also may be belayed safely if a nonspin rappel technique is used: body rappel, or rack or spool or the like, as discussed in the next chapter. If any type of wraparound device is used, or either rope is of laid construction like Goldline, dangerous snarls are indeed likely. Sometimes snarl problems can be reduced by placing the belay station many yards sideward

Three stalagmites crown a huge mountain of breakdown slabs in Indiana's Wyandotte Cave. This historic photo from the Burton Faust Collection was made around 1925 by Russell Trall Neville, "The Cave Man" of the 1920s and 1930s. The angle is upward, so steeply that the flat, level ceiling above the top of Monument Mountain appears tilted.

so that the two ropes form a wide angle. Without special rigging, however, a fall causes a pendulum swing on the belay rope, so this usually is not very practical.

BELAY KNOTS

When on belay, the caver normally ties in with a bowline knot anchored by a half hitch. Those who have taken a jarring fall on a single body loop thereafter use a coil of two, three, or more slack-free loops around their lower chest or a body harness to spread the impact

The bowline is a versatile, virtually slipproof knot that every caver should know intimately. A book on caving is not the place to teach anyone how to coil a rope or tie a bowline the usual way. Those unable to do bowlines, square knots, and the like when numb with fatigue and hypothermia should consult scout handbooks, *Ashley's Book of*

Knots, or the like, practicing over and over until they become automatic. Some cavers thoroughly familiar with the knot prefer simple conversion of a loop slipknot into a bowline. This is done by merely running the free rope end through the loop and "inverting" the knot. This involves a risk of putting the rope end through the loop backward, forming a "false bowline," which is said to slip under a hard pull.

More important is an alternate one-handed technique that can be used when clinging to a rock wall with the other hand (don't ask me why some cavers get into such situations without a sling rope, which would do the job much better). Under such circumstances a belay rope is lowered into the caver's free hand. After he grasps it, the rope handler works it across the trapped caver's back, allowing a few inches' slack. The caver reaches between his body and the rock, bringing his hand and the rope end across the dangling rope. Then he twists his hand down, around the rope, and back up parallel to the hanging part, carrying the free rope end along. This produces a loose loop around the hanging rope and his hand. Next comes the tricky part, best practiced long in advance. With some delicate fingerwork he feeds the rope end through the loop, around the hanging rope, and back into his hand. At this point the body loop is fairly tight, and he has the rudiments of a bowline, but it will not hold until inverted. In order to have enough rope left for an anchoring half hitch, he works several inches of rope through the loose knot. The loop is then inverted by a pull from above and the bowline is complete. The climber can then tie the half hitch, adjust the body loop, and relax. Often the tension of the rope against his body will permit use of both hands as soon as the knot is inverted. If the rope end is passed the wrong way around the standing rope, it produces a false bowline—but that may be better than nothing at this point.

BELAY SIGNALS

During belays, communications often become difficult just when they are most needed. Echoes, natural acoustic baffles, and the innumerable voices of underground water sometimes garble or muffle speech within a few yards. Use of key words permits skilled belaying when only snatches of voices are audible. As a result, extra-audible signals have become standard throughout most of English-speaking North America and considerable regions farther south. Brevity is the keynote:

Climber: *Belay on?*
Belayer (if ready): *Belay on!*
(If not ready): *No!* (often followed by expressive adjectives)

Eventually the belayer is ready and bellows: *Belay on!* Or the climber gets bored and repeats the question. Under no circumstances does he proceed until he hears the confirming phrase. Then:

Climber: *Ready to climb!*
Belayer (if ready): *Climb!*
(In the event of a misunderstanding and he is still not ready, the belayer again hollers *No!*)

After getting the affirmative *Climb!* response, the climber has a choice. Often he chooses to test the rope and the belay position, which is his privilege. First, however, he warns his belayer:

Testing!

Then he leans hard on the rope or simulates a practice fall. If the belayer comes shooting out of his position, they start all over again, with a different belayer, if necessary. This rarely occurs, so he next calls:

Climbing!

The belayer may or may not acknowledge this with the word *Climb!* It's friendly to do so.

At this point there is probably too much rope loose, so the climber yells:

Up rope!

The same phrase is used whenever too much slack is evident.

If the belayer is pulling too hard, the call is:

Slack!

Normally, neither of these receives a vocal response. If the rope is caught, the belayer may respond with *I'm giving you slack* or *Rope caught* or something of the kind.

When the climber needs the rope as taut as possible, he yells:

Tension! (never *"up slack"!*)
If he rests, he yells:
Resting!
When rested, he calls:
Ready to climb!
These two signals are normally acknowledged with a short reply
(*Rest!* or *Climb!*), for a little reassurance goes a long way in the
middle of an exhausting climb. If the climber slips or thinks that a fall
may be imminent, he warns the belayer to get ready:
Falling!
The calm response *Fall!* helps more than might be believed by some-
one who has never been there.

When the climber is in a secure place and has untied the belay rope,
he calls:
Off belay!
The belayer acknowledges:
Belay off!
relaxes and prepares to throw the rope down to the next climber:
Rope coming down!
Except in a serious emergency, all signals are exclusively by belayer
and climber. One exception:
Rock! Rock! Rock! Rock! Rock! Rock!!!!!
Anyone glimpsing any falling object screams this dire word at the top
of his lungs without waiting to see if it really is a rock. No matter if he
is feeling foolish because it was a false alarm a moment earlier. If it
really is a rock or anything else solid, a tenth of a second may save a
hole in someone's hardhat, scalp, skull, and brain.

In particularly difficult situations, even these standard signals may
be inadequate. Wordless voice signals have evolved in some regions.
Yo! and *Bo!* carry especially well. In Europe, whistle signals are
common, but they are not used widely on this continent—I don't know
why, as they seem an excellent idea. Unfortunately, such systems are
far from uniform. The European code cannot be used here because it
conflicts with the North American convention that three signals of any
kind indicate an emergency.

The most logical and reliable North American system seems to be the Huntsville code, in which single signals of any kind mean *Go!*, *Yes!*, *Up!*, or *Out!* Two of anything mean *No!*, *Stop!*, *Down!*, or *In!* Three signals call for whatever emergency action is most likely needed.

Some have attempted to expand this code, but this becomes dangerously confusing to those not specially trained in its language. Every party should agree in advance on its use of the short Huntsville code or some other.

OTHER BELAY COMMUNICATIONS

Telephones solve many such problems, but create so many of their own that they are rarely used on this continent. According to legend, good cavemanship requires all-out efforts to cajole, entice, or threaten someone else into the oft-backbreaking job of carrying the equipment—not to mention laying the wire where it won't get broken underfoot or snarl the ropes and so on. Actually, excellent lightweight models are easily hooked to small batteries and may well be seen increasingly in the future.

"Walkie-talkie" citizens' band radios are currently used much more widely than telephones. These are limited by short range and near restriction to line-of-sight situations. Even cheap 100-milliwatt units can be surprisingly helpful, however. After a recent painful injury at the bottom of the seventy-foot entrance drop of Montana's Bighorn Cavern, voice communications failed, but two toy sets worked fine. More powerful sets require licenses in the United States, Canada, Mexico, and probably elsewhere, but in most areas these are easily obtained by responsible cavers.

Hand-held "ham" units with telescoping $\frac{1}{4}$ wave antennas on a 2-meter frequency and even 4-watt CB units may provide communications at least as much as 4,000 feet into some caves, despite short zones of radio silence. When used near Idaho's deep Papoose Cave, at the focus of an arched shelter cave, parabolic focusing of radio waves causes a mere 100-milliwatt unit to be received at the roadside base camp, nearly half a mile beyond the canyon rim.

Theoretically, the ideal radio for cave use is the voice-activated type, which leaves both hands free for belaying or climbing, yet avoids unnecessary battery drain. These are not widely available on the CB market, however.

Even beyond the limit of voice transmission and reception, clicking the short Huntsville code can be effective.

It hardly seems necessary to add that all radio sets must be: (1) turned on and (2) tuned to the same frequency. Yet fatigue and early hypothermia cause strange lapses underground. A mental checklist helps.

No continent-wide cavers' CB channel has evolved, but a start has been made by a group of radio-wise Missouri cavers who have agreed to use Channel 10. Channel 9 is reserved for emergency use, underground as on the surface. Cavers with two-channel CB units may prefer to choose these two if local traffic on Channel 10 is not excessive. For those with FCC amateur licenses, the National Cave Rescue Commission recommends use of the unofficial mountain rescue frequency during cave rescues: 155.16 MHz.

Israel Putnam would be impressed.

7

Vertical Caving—Descending

In the 1940s when I began caving, each pit was a major obstacle. At best they forced long delays, while each caver laboriously climbed down, usually on belay.

If the pit was unclimbable, we had to go somewhere for cumbersome gear. We were often worn out, before we got back to the cave, by carrying heavy ladders. Rigging with a block and tackle was even more tedious. Mountaineers were rappelling, sliding down ropes on steep mountainsides, but that was in daylight. In caves it was unsafe.

Or so we thought until Bill Cuddington taught us differently, as I related in *Depths of the Earth*. Thirty years later, North American cavers annually log more than a million vertical man-feet of cave. A near cult has evolved, its devotees seeking ever-faster racing rigs in pursuit of an elusive thirty-second barrier for hundred-foot standing rope ascents first broken by Bill Stone. Throughout a continent they seek black chasms twice as deep as the once-legendary thousand-foot pit—a dream that came true at parrot-swirled Sotano de las Golondrinas. Even that remarkable conquest is now merely grueling routine, its record depth already far superseded, the thousand-foot rope ascent often accomplished in less than an hour.

Although he had lots of help, "Vertical Bill" Cuddington truly changed the face of caving.

Depth Estimation

Estimation of pit depths involves much more than the simple gravity formula $D=16t^2$. Important also are the air temperature, the altitude, and the terminal velocity of the rock tossed into the pit. The following table is based on a published analysis by Fred Wefer, though some have criticized his figures as being too high. Gary

Schaechter has obtained considerably lower figures using the formula $D=89.7t-156$ (10 seconds equals 740 feet, for example), and those correspond more nearly with some field observations. The Wefer figures, however, appear to provide the vertical caver with a valuable margin of safety:

SECONDS UNTIL SOUND HEARD	DEPTH IN FEET	SECONDS UNTIL SOUND HEARD	DEPTH IN FEET
1.0	15	6.5	495–545
1.5	35	7.0	555–625
2.0	60	7.5	610–700
2.5	95	8.0	685–790
3.0	130	8.5	750–865
3.5	175	9.0	815–955
4.0	215–225	9.5	880–1050
4.5	255–280	10.0	955–1145
5.0	310–345	11.0	1100–1325
5.5	355–405	12.0	1250–1525
6.0	430–475	15.0	1700–2100

The lower figures are for small, light rocks, the higher for those which are larger, rounder, and denser. They assume a stopwatch, for most cavers count too rapidly even when deliberately slowing themselves. Chanting "one hundred thousand and one, one hundred thousand and two," and so on, may help.

This traditional method of probing pits is not without hazard. Some pit caves have a second entrance fifty or five hundred feet lower. Other cavers might be looking upward at the moment the rock goes down. A test run with a scattering of pebbles is worthwhile.

Methods of Descent

According to one irreverent school of thought, getting down vertical pitches and pits is not really a problem. And there is a particle of truth to such funning, for stopping is the problem, not going.

In the pre-Cuddington days, we tried many kinds of descents. But except for those with something special like the tow truck used to investigate the Earth Cracks (fissure caves south of the Grand Canyon) in the 1930s, or a squad of willing friends to lower and hoist, the choices were rather limited: jumping, chimneying, sliding, climbing down, or using ladders or a block and tackle.

Even today, none of these is quite obsolete.

JUMPING

Jumping is often the most tempting and usually the most easily dismissed of these approaches. Even on belay, a caver loses control of the situation when he jumps. We all do it sometimes, at the cost of occasional sprained ankles and assorted bruises and abrasions, but no farther than a foot or two. The Inevitable Law of Inverse Perversity seems to become operational at about eighteen inches. Nice solid dirt floors turn out to be under half an inch of slippery glare ice, or a transparent pool. A dirt-covered "false floor" canopy a quarter-inch thick has quite a bit of nothingness beneath it. Or the big square rock that looked so solid was delicately balanced on a narrow, hidden pinnacle. There is almost always a better way than jumping.

SLIDING

Sliding isn't much better than jumping. Where there is a nice safe place to stop at the bottom of a snow, mud, gravel, or guano slide, sliding may be safe, but the slider must expect surprises. Jolting down, toes or heels driven deep for balance, is usually much superior.

On the second 1940 trip into the inner reaches of Virginia's historic Blowing Cave, National Speleological Society fathers Bill Stephenson and J. S. Petrie came to just such a mud slope. Pete had picked his way sloppily to its bottom in the initial breakthrough and knew there was a safe run-out spot below. So he happily launched himself. Since his last visit, however, the cave's water level had risen fifteen feet. He never touched bottom.

In glacier caves, spelunkers normally have mountaineers' ice axes and the knowledge of how to use them on snowbanks without impaling anyone. Once or twice on mud slopes, I've wished for one, or a fixed rope, for balance.

As for sliding down a rope, I've done this once in the last ten years, on my *second* time down an exceptionally tight chimney in the Nicholson's Pit area of Carlsbad Cavern. Probably 99 percent of the cavers sliding down a rope should be using it for a belay or rappel instead.

CHIMNEYING

Chimneying is much more than a method of descending (or ascending) in round, chimneylike domepits, moulins, and the like. It is often the easiest way to get around underground. Many look first to chimneying wherever it is possible to wedge oneself in place by counterpressure against two walls, or even the irregularities in one wall. One can chimney up, down, or lengthwise in long, narrow cracks, thus staying dry in stream slots, or high up in canyon passages that narrow uncomfortably at the bottom.

Chimneyable routes turn up in unexpected places: a shallow waterfall slot on a vertical wall, or the corner of a large, steep-walled

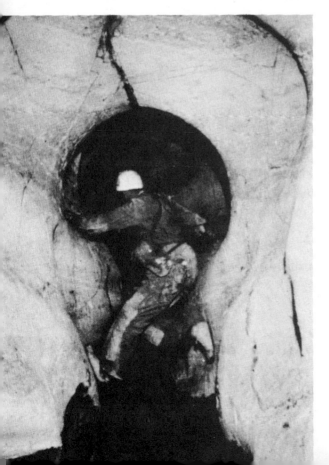

Spelunker chimneying in Jewel Cave, South Dakota. The rounded contours of the passage reveal its phreatic origin. The joint that permitted its development is visible in the ceiling. A thin speleothemic coating stops abruptly just below the caver's left foot. Such features permit "reading" much of the past history of each cave.

chamber. Before unpacking ropes, take a quick look around for an easy chimney route.

No matter what the direction, the basic principle of chimneying is counterforce. In practice, this combines with small but convenient footholds and handholds—wedge holds, clutch holds, grasp holds, squeeze or pull holds, anything. Not infrequently, excellent routes permit cavers to zoom happily up or down or along a narrow passage spiderwise, one arm and one leg to each side. Perhaps more often, one or both soles must be planted on one side while the explorer wriggles his back up or down the other. Or his hands are pressed against one side while his feet or knees move the caver along the other.

Cavers of different lengths and diameters rarely use identical techniques in any given chimney. Some may need a belay because the crack that delights their fellows is too narrow or too wide for their particular configuration.

Before beginning a chimney descent, thought must be given to the return. For some, the ascent may be excessive. Too, a very tight chimney may require special rigging. Or, if a slip would cause no harm, this may be one of the extremely rare situations where a standing rope may properly be used for direct aid in the ascent. A self-belay (see page 186) is desirable.

One special precaution: chimneying against a thin blade or flake of rock, ice, or even snow may cause collapse of much more of the cave than would seem likely.

DESCENDING ON LADDERS

Since almost all cavers now know how to rappel, rope and cable ladders are now largely ascending devices. When rigging a ladder, cavers obviously have a rope along for belaying, so each might as well rappel; it's usually easier and probably safer. In the rare situations where someone chooses to climb down a ladder, all that needs to be emphasized here is that, if the ladder is hanging free, the heels be inserted rather than the toes. One should stay as straight up and down as possible. Be belayed. Always.

For the ascent, a figure-eight chest loop or harness is needed, so it might as well be worn on the descent, too.

Climbing Down

Climbing downward is not at all simple. Indeed, it is much more difficult than climbing upward, in caves as everywhere else. Powerful muscles drive the human frame upward and hold it in place. None is specifically designed to hold back against the pull of gravity.

Even among cavers who don't yet need bifocals, the finding of a downward route often requires dangerously ludicrous shifts of balance, for the eyes are annoyingly far from the route. Every lurch signals momentary loss of balance. If the descent is at all tricky, lurching is an indication for a belay. Mountaineers and hikers talk and write about grace and rhythm. These are a consequence of making one's momentum work to his advantage, and should be a part of the descent of rock piles, steep slopes, and rock walls. But this is much more easily learned and accomplished when upward bound.

As a cave steepens, cavers shift from scrambling to three-point suspension, and, from facing outward, turn sideways, then increasingly face the cliff. Thus we remain in place when a rock rolls underfoot, or comes along as an unexpected memento. With still more steepening, individual holds become increasingly important: small horizontal flats are convenient for the arms while seeking footholds below, and a narrow crack into which a fist can be wedged may be welcome indeed. On slippery slopes the best stance is upright. The seeming security of leaning inward, close to the comforting wall, proves false as the feet go out from under.

Avoiding rotten-looking or loose-looking rock and ice is common sense. Perhaps not quite so obvious is double-checking handholds, to be sure that they are more than an insecure wedge about to detach from the wall.

"Jam holds" with a knee, elbow, or foot in a crack are especially treacherous during descents. The jammed member tends to stay behind when the rest of the caver moves on.

Rappelling

Hanging in midair on the end of the belay rope is more than slightly unpleasant. At the point where climbing down becomes difficult and dangerous, rappelling is becoming increasingly easy and safe.

The caver begins with the enormous advantage of being able to rig rappel ropes and ladders, and should make the most of it. Yet nobody wants to spend time and energy needlessly uncoiling and recoiling ropes and ladders at every little rock pile and ledge. The quicker and easier choice is not necessarily the safer, and the safer choice for one team may not be so for another. As a general rule, however, rappelling is preferable whenever it will save time and energy.

Originally a comparatively minor mountaineers' descent method, subterranean rappelling is diverging farther and farther from its surface counterpart. It is an unforgiving technique, with fewer fail-safe precautions possible than many would like. Severe, even fatal injuries in seemingly easy locations increasingly demonstrate overconfidence and undertraining. Yet rappelling is the key to exploration of deep caves and pits.

BODY RAPPELLING

In the old pre-Cuddington days, rappelling consisted of wrapping a standing rope around the caver's body, leaning backward over the lip of a pit, and walking, bounding, or sliding downward. The speed of the descent was controlled by the friction of the rope around the body. And, as everyone knows, friction means heat.

Felt-padded leather rappel pads flourished briefly, but have almost vanished as newer techniques displaced body rappelling—fortunately, for such pads encouraged overconfidence. In chill caves that require heavy clothing, body rappels are still seen for drops of less than fifty or seventy-five feet. In some ways, this is a safer technique than some mechanical devices mentioned below. It is almost impossible to get caught in midair by one's shirttail, for example. But it is fairly easy to fall out of a body rappel by poor feeding of the rope and getting the legs too high. And the rope burns come at inconvenient locations.

Although its use is more and more limited, for emergency use every caver should know the body rappel. To begin, the spelunker straddles a standing rope at the edge ("lip") of a drop, facing the anchor. With one hand (say his right) he grasps the rope behind him, sets it under that hip (the right) and brings it forward across his side (the right) and chest (the front) to the opposite shoulder (in this case, the left). Thence it passes diagonally across his back into the hand he started with.

In this case, the right will be the control hand. It will feed the rope upward if it does not slide freely under his weight, or steer it across a greater body surface if he needs to slow the descent.

THE SELF-BELAY

The other hand has several functions: balancing, fending off sharp rocks, and the like. Plus one major, highly specific use—management of the self-belay loop or device.

The self-belay in rappelling is a special use of the Prusik knot or Jumar or other ascender described in the next chapter. These slide freely along a rope when open, but grab and hold under tension when released. Although they are quicker on some ropes than others, and not ideal when a rope is icy or slimy, they work pretty darn well.

The Prusik knot is fashioned on a small rope or sling loop of webbing that fits around the upper chest or fastens to a chest or seat harness. An ascent device would be anchored rather similarly. If the rappeller faints or is hit on the head by a falling rock, either will probably catch him in midair. An interesting situation faces him when he regains consciousness, but at least he's alive. Rappellers desiring to avoid an uncomfortable spell hanging in midair have in a pocket at least one additional ascent sling and preferably two, because the self-belay Prusik often jams so tightly it must be cut loose. This does not necessarily mean that a Jumar or other ascender is automatically better here. Without extra precautions, these are likely to abrade the rope. Only time will determine which is the best.

If an ascent device is used, it must be held wide open throughout the descent; if a Prusik, the noncontrol or guide hand must keep it open and feed it downward, not quite at the point of grabbing. The hand must *not* ride on the knot. If the rappeller loses control and freezes with his hand on a loose knot, it cannot tighten, and his fall will be unchecked. This must be specifically learned, for it is contrary to the normal survival instinct: grasping the rope though it be cutting through skin, tendon, and bone. Practicing over and over on the surface is essential. And a slow rappel is more proper anyway.

MECHANICAL RAPPELS

THE CARABINER WRAP

The carabiner wrap is the simplest mechanical rappel technique, and probably the riskiest. The braking friction is often not quite adequate, requiring additional body friction as the caver descends. This is obtained by passing the rope over the hip, around the leg, around the back, or by bending it upward—this usually a last resort. Shirttails, loose rope ends—the damnedest things seem to get caught in the rig. It is particularly hard on braided kernmantel ropes, sometimes even twisting the cover away from the core. Laid manila and nylon have been known to shred. Even well-known experts have had ropes snake out of carabiners they swear were in the proper position—the carabiner gate facing up and opening away from the body. This is supposedly prevented by wrapping the upper (fixed) end of the rope around the carabiner arm (side opposite the gate) *twice,* but many experts strongly recommend against use of this technique except in dire emergency.

Rigging in and stepping back into the friendly, familiar darkness is so easy with a carabiner wrap, however, that it is probably here to stay for drops up to about fifty feet. It often functions well, especially against a wall.

Or so it seems. Midair spin becomes a problem, even on a fifty-foot descent. And below the rappeller the dangling rope kinks and dances like some weirdly maddened serpent, walking about bizarrely like a whirling top, occasionally snarling fast. So far no one is known to have been trapped underground in this way, but cavers' luck runs out sometimes.

When a carabiner wrap is unavoidable, only an adequate locking carabiner with a screw "safety" gate should be used. Many older types have dangerously weak gates, but newly tested models are now available from reliable outlets like Bluewater and Mountain Safety Research.

CARABINER–BRAKE-BAR RIGS

By passing the rappel rope over brake bars inserted onto two carabiners in series, the spin problem is virtually eliminated. The rope is much less stressed, and control is much better. The threaded nut can be removed from some locking carabiners, thus permitting their use with brake bars. Yet other problems exist.

Single-carabiner–brake-bar rigs are occasionally seen, despite several near tragedies from serious equipment failures. A single brake bar tends to unlay a rope, so two are normally used even with a single carabiner. Each is positioned so that the weight of the rappeller holds the rope against the bar, and the bar against the arm of the carabiner. Uneven, bouncy descents may momentarily release this pressure, and ropes have been known to snake around the bars. Other problems with this rig occur at the lips of drops, where pressure may cause the bar to pop off or "ping" away. The gates also undergo particular stresses here, and the second carabiner is such cheap life insurance that the single rig probably should never be used. Rope kinks, knots, sudden stops, and rapid rappels are particular problems with this method. In 1972 testing showed that, during a rapid rappel, the brake-bar temperature exceeded 250° F. with this rig. Nevertheless, it is generally satisfactory for drops up to about two hundred feet.

THE RAPPEL RACK

When forty pounds of rope dangle below a rappeller, several curves of the rope around any device produce so much friction that the rope often has to be hand-fed upward. Yet near the bottom, with little

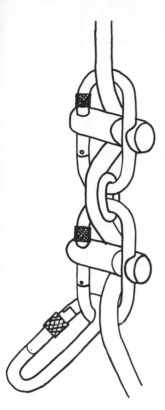

Fig. 2. Two carabiner–brake-bar rappel rig. In this particular rig, carabiners are linked by D-ring. Carabiner hooking rig to Swiss seat has not yet been locked.

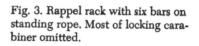

Fig. 3. Rappel rack with six bars on standing rope. Most of locking carabiner omitted.

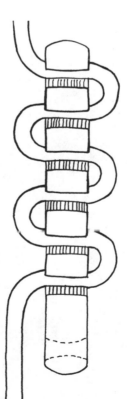

Fig. 4. Figure-eight descender on standing rope. Locking carabiner and harness omitted.

rope weight remaining—and with little warning—friction may become dangerously scant. For this reason, a third carabiner and a variety of other devices were investigated: the Pompier hook, oversized carabiners, the Cuddington rappel spool, the Rappel Hammer, the Allain descending hook, Patten's Hook, and many more. Several provided especially good heat dissipation, but many threatened escape of the rope in midair. The best provided incremental control by permitting a caver to add or remove a turn or two of rope—always a bit tricky in midair, and usually at the cost of additional spin. Some required the repeated lifting of forty pounds of rope overhead or other extremely tiring rope adjustments during a thousand-foot rappel. The three-carabiner–brake-bar system worked much better, and led in turn to such specific incremental control devices as J-bars, brake-bar ladders, whaletails, and especially the rappel rack.

Invented by John Cole, the rappel rack today leads the way in safe descents of a thousand feet and more. Some believe that rope weight will limit use of the standard-length rack at fifteen hundred to two thousand feet. In deeper pits, cavers may have to climb down standing ropes, perhaps "thumbing" ascenders, as described in the next chapter. So far, however, such pits remain dreams. For long mechanical rap-

Harry White "rigged in" at the lip of the 1,345-foot drop of El Sotano, Queretaro, Mexico. Note rack, Jumar, pack, canteen, and other gear.

pels, the rack is widely acclaimed as the best all-around descent device yet developed.

The rappel rack is basically a U-shaped holder for five or more brake bars, with a secure loop at one end to snap it to a Swiss seat or other harness, and a screw nut on the other to keep the brake bars from sliding off. Some homemade models still are seen—including so-called Super-Racks with heavy machined bars for greater heat dissipation—but most American vertical cavers today use PMI, Bluewater, or some other reputable commercial model. Bill Cuddington's favorite is fashioned of "1018" metallurgist-approved steel rod, heated and bent with extreme precision. Others claim an advantage for cold-bent steel, even though such cold-rolled steel must be kept lightly oiled when stored. Stainless steel models must be washed in kerosene.

Some models have a free space beyond the brake bars. This permits the rappeller to "lock" the hanging rope, looping it through the arms of the rack. Vertical cavers using other types say they don't need this feature.

When beginning a long rappel, the caver normally starts over the lip with the rope laced through one more bar than appears necessary, even if he must hand-feed the rope upward through five or six bars. Various conditions at the start cause enough unpredictability in the friction that only one bar is removed at a time.

As the descent progresses, more bars are added as needed. Additional friction can be obtained by jamming the bars together. Misuse of the bars can exhaust an unskilled caver, but comfortable adjustments are usually easy. With practice, virtual fingertip control can provide an almost effortless descent.

Five-bar and six-bar lengths ("short" and "regular") are widely available. Some racks are designed to be used only with specially designed brake bars but most accept standard types (see below).

BRAKE BARS

Both hollow steel and solid aluminum brake bars are available at many caving and mountaineering stores. The aluminum bars dissipate the heat of rappels better than those of steel, but wear rapidly (especially on gritty ropes) and leave aluminum streaks along the rope. Only seamless-tubing types of steel bars should be used. Both stainless steel and cold-drawn carbon steel models are available; the same precautions apply as for racks.

OTHER CURRENT RAPPEL DEVICES

The whaletail is a solid aluminum descender with a series of grooves along one side through which the standing rope is threaded. A safety gate controls the two uppermost slots. Ropes occasionally creep out of the lower slots. While the device is sold in the United States, it has never become as popular here as in Australia. Similarly the Petzl bobbin (widely used in France) and the figure-eight descender (primarily an English device) have not become popular here. The latter is simpler and perhaps safer than carabiner wraps and merits further consideration for short drops.

RAPPEL EMERGENCIES

Short rappels are commonly accompanied by happy chatter. Longer drops involve communications problems like those mentioned in the last chapter. In some parts of North America, any voice contact during a rappel is considered to mean an emergency—usually loss of control by the rappeller. Fellow cavers already on the bottom are expected to respond instantly by throwing their weight onto the rope. This increases the rappeller's friction and slows or halts his descent.

This is a grave hazard for all concerned. The rappeller faces violent midair stresses; and those below, whatever is loose in his pockets and pack. Nobody seems to have measured the effect of such sudden stops on the rope. Yet, in dire emergencies, it usually works without undue complications. Better communications, however, are an obvious need, for the victim may merely be asking for a pocketknife to cut his shirt-tail loose.

Several other types of rappel problems mentioned earlier in this chapter may necessitate aid through tandem ascents and similar techniques mentioned later in the book. Occasionally, someone will have to rig a parallel standing rope onto which the rappeller can transfer (see Chapter 8).

RAPPEL SPEEDS

The first caver to descend needs to proceed slowly to clear loose rocks wherever he can reach them. Even for the others, a slow, steady descent is preferable, except in waterfalls and similarly nasty spots where speed is vital. In skilled hands, the load on the rappel rope may be no more than 20 percent greater than the weight of the rappeller's body alone.

This is a dramatic reversal from the old learning days, which saw showy "jump rappels," with the caver swinging far out into space with each magnificent burst of action. Such bounces were a bit hard on the rappeller's body and hands—but even harder on the rope, the anchor points, and anyone struck by dislodged rocks.

Some rappellers still traditionally z-z-z-zip down the rope at a rapid, steady pace, then brake as the bottom nears. Even if he brakes smoothly and gradually, however, the rope may overheat, the anchor points overstrain. Should something go wrong and the auto-belay grab in midplunge, the impact is especially severe.

RAPPEL STANCES

In earlier days, the best rappel stance was mistakenly thought to be at a sharp angle, leaning far back as if the rope were the far wall of a chimney. And indeed, if the descent is far from vertical, this may produce better traction. On near-vertical and free descents, however, the upright position is essential. Mountaineers speak of "playing" off the wall for better control. From free rappelling, however, cavers have learned that, just as in ladder climbing, the center of balance should be as straight up and down as possible. At the lip of the pitch, it may be necessary to lean back to get started. But as the rappeller reaches the point where he can be comfortable in the fall line, he should be vertical, toeing the wall (where he can) only as needed to control spin. Especially in this position, whipping the hanging rope around the body will almost always bring him to a stop.

DON'T BURN THE ROPE!

At least once, a rappeller has burned his rappel rope in two. One well-known vertical caver sports a horizontal wire loop on his helmet, "just to be sure." Electric cavers have no such problem, and carbide cavers need not. A few surface practices with a flaming carbide lamp usually suffices. But it is a good idea to protect the rope with a length of slit garden hose.

ANCHOR POINTS

Every caver must be comfortable with a wide variety of anchor points, both natural and artificial. Tensionless rigging is preferred but not always possible. Perhaps the nicest is a pair of trees several inches in diameter. Three or four wraps around the nearer tree, then back to the other tree for a bowline on a loop, and a tensionless anchor is set. If only one tree is at hand, a bowline on a coil or a midshipman's knot is desirable. The closer a rock or other object resembles such a tree, the better it serves. Sharp edges and major roughness must obviously be padded.

Sometimes the standing rope is anchored to a carabiner, from which one or more sling ropes pass to the ultimate anchor points. Competent mountaineers calculate that such sling anchors are reasonably safe if they have a capacity of around 600 pounds, after deducting loss of strength by bends in knots and over the carabiner. This, however, is based on high-stretch ropes. If a low-stretch rope like Bluewater II is suddenly stressed by the rope sliding off a projection and the rappeller free-falling a few inches, the sling and anchor receive a more severe stress. For maximum strength, such sling anchors should form as narrow a loop as possible without risking rope damage, and multiple anchor points are often needed. Ideally, the rope should pass over a pulley suspended well above the lip and anchored separately. A high-stretch rope is especially valuable for this pulley anchor, since it will absorb much of the shock of a sudden stress on the standing rope. One or more such anchor slings are easily shortened by a few turns around the arm of a carabiner.

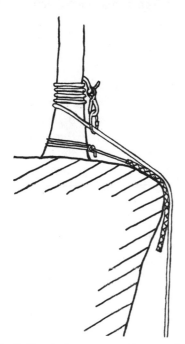

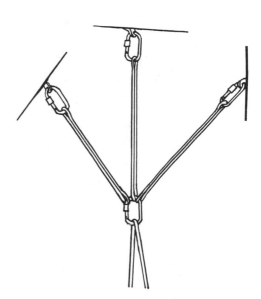

Fig. 5. Tensionless rigging plus use of a rope pad.

Fig. 6. Diagram of three-piton anchor. Omitted for clarity are sling knots and/or stitching, tightening carabiners, and additional safety slings. Rope thimbles (not shown) might well be used.

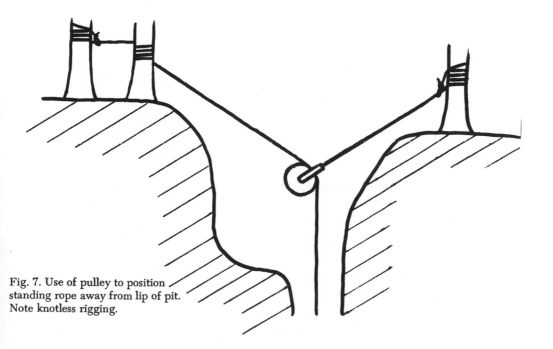

Fig. 7. Use of pulley to position standing rope away from lip of pit. Note knotless rigging.

PROTECTING THE STANDING ROPE

At least the upper part of the rappel should be examined suspiciously, and the safest route chosen. The lip is an obvious danger point, and often needs special padding. Ropes may also rub farther down a pitch. Sad experience has popularized special protective rope pads. One—the Cuddington pad—consists of a canvas rectangle about six feet long and two feet wide. It has a long rope tail that is attached to the rappel rope or a second standing rope by a Prusik knot or other ascender (see Chapter 8). Such pads are primarily used to protect the rappel rope against dangerous abrasion, but also help avoid mud and slime. In the event of deep, soft dirt or snow at the lip, padding alone will probably not keep the weight of the rappeller from burying it deeply. But it is worth trying before undertaking the long, strenuous construction of a firm lip.

Some use slit garden hose to protect rappel ropes from more than carbide lamps. If nothing else can be done to minimize abrasion, it can be averaged along a considerable length by periodic repositioning of the anchor knot.

Abrasion is not the only hazard ropes face during and after rappels. Especially after long handling they acquire a salty taste attractive to animals. And rock-throwing children, rope-stealing adults, and other modern sociological problems are causing more and more vertical cavers to post guards to protect their ropes.

RETREATING ROPES

Experienced cavers are more than a little skeptical about the lengths of their ropes. Every year some happy vertical type descending on high-stretch nylon forgets about the stretch until he lands and lets go: Presto! Out of reach! Almost as embarrassing is the unexpected ascent of climbing slings attached to a high-stretch rope as soon as one climber is out of the way. As he ascends, so do they—not nearly as far, of course, but far enough for a horselaugh and considerable annoyance.

ROPELESS RAPPELLING AND ITS PREVENTION

If a rappel rope doesn't reach the bottom, a knot at the end reduces the incidence of air or ropeless rappelling, a distinctly undesirable technique. A locking carabiner clipped into the knot automatically comes into the braking hand, and is easily snapped into the similar carabiner of the Swiss seat or harness.

PITONS AND OTHER CLIMBING HARDWARE

Artificial anchors are rarely as satisfactory as natural ones. Topflight rock climbers assert that no caver knows how to use pitons, expansion bolts, or other climbing hardware properly—and they are not far wrong, for piton mastery requires years of experience on "rock climbers' rock," on rotten stuff that barely makes mountains, on ice, on firn, snow, sod, and other annoyances. In cavers' hands they fail often enough that no one should use them underground without considerable surface practice—and not at all if safer techniques will suffice.

Most cavers know that pitons are long, thin wedges of metal with a ring or hole at the blunt end for a carabiner. Some are flat; others—angle pitons—are bent into a V on cross section. Among the flat types, so-called vertical pitons have the ring in the same plane as the blade, while "horizontal" types have it at right angles. They are fabricated of a variety of metals. Rock climbers have largely abandoned old-style malleable iron types in favor of special steel alloys, which include chromium and molybdenum. Cavers likewise would do well to use these "chrome-moly" types.

Pitons are hammered into cracks until a clear, ringing note indicates a contact so excellent that they will withstand a considerable shock. Then they are tested by yanking back and forth to see if they come loose readily. If nothing happens, a carabiner is attached and the test repeated. The gate should be away from the wall or rock so that neither banging nor a fall is likely to open it. Given the proper piton and proper techniques, even vertical or downward-slanting cracks can be used. Innumerable shapes and sizes of pitons are manufactured, since few cracks come in standard sizes. Rock climbers may carry

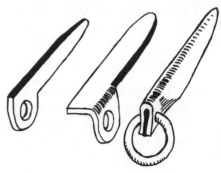

Fig. 9. Safe and unsafe methods of insertion of angle pitons. Three-point contact is necessary, whatever the angle of the crack into which the piton is driven.

Fig. 8. Basic types of pitons used in caving: vertical, horizontal, and angle. Ring pitons are being used less.

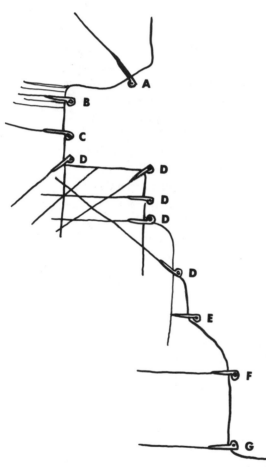

Fig. 10. Good and bad piton placement: *A*. Not ideal, but safer than it looks if well-placed. *B*. Dangerous. Beds are too thin. *C*. Probably good, although projection below must be padded, and some risk because of closeness to thin beds. *D*. Dangerous because of badly fractured rock. *E*. Even more dangerous because joint ends. *F*. Good, but piton not adequately driven into crack. *G*. Good, but projection below piton may require a sling. *H*. Dangerous.

dozens, but cavers with some training in their craft often prefer to limit themselves to two or three basic types: perhaps one horizontal piton with a wide blade, one vertical, and one angle type; perhaps a few others until weight becomes a problem. Obviously, all are valueless without a hammer. Not so obvious is the need for a thong snapped to the piton to catch it when it gets loose.

If their use cannot be avoided, they should be placed so that a fall will jam them more deeply, or as close to this as possible. In choosing a crack, consideration must be given to its widening, deepening, and lengthening by the thrust of the piton, perhaps splitting off a block of rock. Good piton cracks are scarce in caves, and no piton found in place can be trusted, for cracks widen faster in many caves than on the surface.

Even in broad daylight, in the hands of experts, too many practice falls are accompanied by an ominous *poing! poing! poing!* as a series of pitons pulls out under an impact each was supposed to withstand. Where used as an anchor point, one or two additional back-up pitons in different cracks and/or expansion bolts provide at least moral support.

Ordinary pitons can rarely be used in lava tube, glacier, and some other types of caves. In friable rock, ice pitons are sometimes worth a try. These can sometimes be used in the walls of glacier caves, but rarely in glare ice; ice screws may be preferable.

CHOCKS

Aesthetically chocks are far superior to pitons. They are artificial chockstones wedged in cracks and 'jammed" tight. They are directional safeguards, intended to hold only when force is applied in a single direction (normally down), and their safe placement can be even more of an art than that of pitons. Only beginning to be seen in North American caving, initial reports on them suggest that suitable cracks may be more common in alpine caves and in dense limestones. Removal is usually much easier than that of pitons, sometimes too easy, in fact. Increased safety results from their use in pairs or groups designed to function together in case of misdirected forces.

The simplest and earliest chocks were small rocks around which webbing was passed. Lengths of pipe have also long been wedged in narrow cracks and chimneys. Commercial models now provide a considerable variety in forms, materials, and malleability.

EXPANSION BOLTS

Considering the exhausting, maddening delays needed to emplace them, expansion bolts are used more commonly than might be expected. Several rather dissimilar types are available, each requiring interminable hammering and each coming with specific directions, which need not be repeated here. The Phillips bolt drills its own hole as it is hammered into the bedrock. These have a sleeve that fixes the bolt as it is screwed into place. Unfortunately, these tend to become dull before entering hard limestone full length.

In soft limestone, sandstone, or gypsum, the Star Dryvin—a nail-like object three inches long, readily driven into the sleeve in the drilled hole—is especially easy to use. Somewhat similar but stronger, the WEJ-IT bolt is especially popular around Huntsville, Alabama.

For the comparatively few hard, crystalline limestones that do not flake, the Rawldrive is especially useful. It has a flanged central section that squeezes together when driven into the drilled hole. A bolt one-quarter inch in diameter and one and a half inches long is generally satisfactory; use ⅜-inch bolts for high-load rigging.

When the hole has been drilled and the bolt placed satisfactorily, a special metal hanger is attached to the bolt, and a carabiner attached in turn. The hanger must be flat against the wall and the unit positioned to resist a fall. Flat hangers must be positioned with particular care, lest a fall apply leverage against the bolt.

Extra equipment is routine. Many cavernous limestones are so hard that the kit must include several drills, a drill holder, a piece of hollow rubber tube for getting dust out of the hole every few minutes, a whetstone, a hammer, and plenty of patience. Even with this, more than an hour is usually needed to install each bolt in hard limestone.

This entire section is intended to imply that I rarely trust pitons in my own hands and those of most cavers. Others share this attitude.

Bill Cuddington stresses that cavers should never rig to a single piton or bolt: "Always back everything up." Preferably the back-up forms a V-shaped anchor of three pitons and/or bolts. The extra rigging time is valuable life insurance.

FLUKES, PICKETS, AND OTHER ANCHORS

Occasionally no satisfactory projection, piton crack, or other passable anchor point can be found near or transported to the lip of a pit or moulin. Under these circumstances, the best approach is going home and forgetting the whole thing. If this is not feasible, such problems may be approachable by the use of special mountaineers' tools, like snow flukes, pickets, or snow pitons. The latter are one-inch aluminum tubes or angle irons, driven three or four feet into snow or dirt, at an appropriate angle. Normally they are used in pairs or lines, one anchoring the next. Ideally they are driven into the reverse side of

An old postcard view shows one method of vertical caving!

a crest or lip, with the rope passing over something that keeps it from burying itself under tension. If the descent is not absolutely in the fall line, a V of pickets or special pitons should be pointed in the direction of fall.

In hard ice, a series of ice screws can be used if the ice does not fracture. All such techniques, however, require accepting a much greater risk than is normal for caving.

Other Methods of Descent

On rare occasions, other methods of descent are a happy choice. The use of winches and blocks and tackle, for example, may be essential for safe descents in icy waterfalls. Faced with the task of running an entire national convention up and down the 185-foot entrance shaft of the Devil's Sinkhole, resourceful Texas cavers turned to automotive power. Rigging a parachute harness to a boom and pulley, they marked out a set course and ran a car back and forth all day, lowering and raising 285 cavers without incident. Less careful drivers seeking to duplicate their technique have dropped some loads harder than planned, at the bottom.

Which is really the problem on all descents. Beware the sudden stop.

8

Vertical Caving—Ascending

Descents are mostly fun, especially on standing ropes. Getting back up under one's own power is darned hard work, but it feels good after it's all over (somewhat like beating one's head against a wall).

Here, too, mountaineers politely sneer at cavers' skills, asserting that cavers are lousy climbers (upward, they mean). Usually they're correct, for the challenges of caving and climbing are dissimilar—hence the responses. Few mountaineers ever face the need to hasten upward through narrow cracks drenched by chill torrents, to fetch help for a friend, in shock and with a crushed foot, six hundred feet down in the dark. When confronted with a rock wall or steep slope that must be scaled anew, few cavers can match the grace and agility of even the average rock climber. Yet, with the development of new ascent techniques and devices, many cave men and a few cave women can ascend a hundred-foot overhang on a standing rope in a minute or two, as occasionally we must. A new breed of vertical caver has suddenly evolved—at home in midair, uneasy on the not-quite-sheer rock that delights climbers.

The need for speed is obvious in the icy subterranean waterfalls of Montana and in emergencies everywhere. In other cases the need is more subtle. A significant energy loss occurs even when sitting comfortably in a Swiss seat, clinging to a cramped ladder, or balancing on a worrisome, crumbling ledge, draining the caver's reserves even more. Overfast ascents present an obvious problem, but modern vertical cavers have reversed the ideas we once held. Today we ascend rapidly, we rappel slowly.

Subterranean Rock Climbing

New standing rope techniques notwithstanding, the basic principles of rock climbing are important to every caver. The finest belay cannot compensate for poor climbing technique.

Especially important is the idea of climbing with the eyes before climbing with the feet. In climbing or clambering or hiking, on rock or mud or whatever, the easy way is the best. Skilled cavers are on the lookout constantly for an easier route.

Momentum is equally important. It carries the caver onward and upward—rhythmically, smoothly, testing each foothold before committing his weight. To make it easy for his legs and his belayer, he keeps upright, his center of gravity neatly balanced. Beginners feel safer leaning inward, as if they could grasp the rock to their breast, but in that position they cannot see as well nor move as well. Worse, leaning pushes the feet outward. On loose rock, mud, or ice the feet slip, and on steep rock the awkward angle may chip off toeholds (Not that a caver or climber often puts his toes in "toeholds." The edge of the foot usually works much better). A relaxed upright stance compresses dirt, mud, snow, and even loose rock. Only a little, perhaps, but a little often suffices.

Wherever possible, the legs should do all the work. Few realize how weak are the most powerful arms in comparison to the muscles of the lower extremities. As the ascent or traverse becomes more difficult, the caver may begin to use his hands, but for balance and security rather than direct aid. When risk becomes significant, three-point suspension becomes the rule, and only one limb is moved at a time. A smooth, continuous flow of motion helps avoid overcommitment to any single point. With the knees locked and the body vertical, momentary pauses can be worked into a surprisingly restful rhythm.

Upward steps should rarely be as high as the knee, and handholds should be close together, preferably at chest level. Where only tiny footholds are available, their use as toeholds is much more tiring than using the side of the foot would be. Some can be toed momentarily while moving smoothly to a better point beyond. Spread-eagling and leg crossing are disastrous. Especially if secure handholds are present, the feet often can be shuffled on almost imaginary ledges. Kneeholds should be avoided whenever possible.

The hands serve primarily to balance the caver, to test the rock, to move dangerously loose slabs, and the like. Not every loose rock should be removed, for it may be safer than whatever is beneath it. If uncertain, it should not be pulled outward. Rather, it should be tested with increasing stress in the direction of force that will be used in the ascent.

At times the hands must be used for direct aid, but rarely for pulling the caver upward or pushing him up a short pitch. Cling holds may face up or down or sideways; rarely can they be used above shoulder level. "Undercling holds" can be as low as the waist. At times the hand, fist, or fingers—or both hands, elbows, or even feet—can be jammed effectively into convenient cracks or holes. These are so-called rotation or jam holds. Considered a separate type of handhold in most rock-climbing texts, down pressure of the hands intergrades into the chimneying that "comes natural" to most cavers. Layback techniques are rarely used in caves. For ascents of such difficulty, other spelean techniques are usually available at less risk.

When Joe Caver has to work his way upward into a previously unprobed part of a cave, he wishes he had spent more time learning rock-climbing skills. On first descents, loose rocks harmlessly boom or clatter downward. On first ascents, mere minor boulders may become unconquerable obstacles.

Here the skills of the few caving rock climbers may become irreplaceable. Pitons or expansion bolts may have to be driven up endless walls to protect the leader and to anchor slings or rock climbers' stirrups (very short lengths of cable ladder). Mountain Safety Research auto-belayers may become worth their weight in gold. Rarely, features of tension climbing (discussed at length in mountaineering texts) may be helpful.

The infuriatingly slow conquest of waterfall pitches like those of West Virginia's Overholt Blowing Cave requires wet suits and incredible dedication. Climbing crevasses from the inside should be little harder. Glaciospeleologists are beginning to discuss new uses of ice pitons and screws.

Rope Climbing

Purists insist that no caver should ever hand-climb a rope. I personally learned this the hard way years ago in the legendary Cave of the Winding Stair beneath the friendly Mojave Desert of California. Those interested in how *not* to surmount an overhang may be interested in the account in my *Adventure Is Underground*. Even on belay, hand-over-hand climbing for more than a foot or three must be condemned. In all but a few truly exceptional individuals, the arms give out rapidly and unpredictably.

In tight chimneys and very occasionally in other locations where a fall would cause no harm, a standing rope may serve momentarily as a welcome balance point or third point of suspension. This is wholly different from hand-over-hand climbing, and is so rarely relevant that I debated whether to include it here. But occasionally it is important. If a large party needs to pass such a point rapidly, overhand knots every twelve to eighteen inches can speed it considerably, perhaps the only valid use of a knotted "climbing rope" underground.

Some cavers are said to use loop ladders—dangling, boot-sized loops every foot or so along a standing rope—but if so, no one seems willing to admit that he is among their number.

Rope Ladders

The use of rope ladders and cable ladders on this continent never reached the heights (depths, if you prefer) achieved in France, where giants of caving regularly used them to conquer vertical pitches of more than three hundred feet. The early flowering of vertical caving saw a short period of ladder triumphs here also, but today their use is much more limited. They are far from obsolete, however. In many regions special standing rope ascent techniques are not needed because the vertical pitches are all short. Cavers facing drops no longer than sixty or seventy feet can get along with ladders and competent belayers. At such depths, moreover, ladders can evacuate large numbers of cavers much more rapidly than standing rope techniques.

While the use of rope and cable ladders has been reduced greatly as a result of development of standing rope techniques, they are still useful in many circumstances.

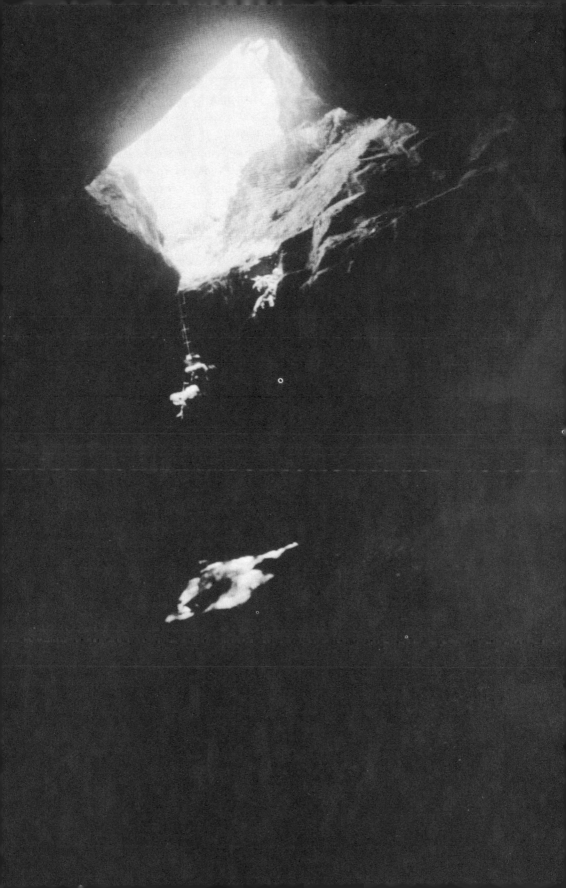

When an emergency develops, it's nice to have a ladder rigged and waiting.

Novices are fun to watch on ladders. If no one has tipped them off, they put their feet on the rungs, toe first. Then they lean back, trying to pull themselves up. Sometimes their feet end up on top.

The key to ladder ascents is keeping the center of balance upright and as close as possible to the ladder—plus pushing the body upward with powerful leg muscles. In free air, the caver climbs around the ladder, putting his heels onto the rungs, not his toes. Against rock walls, the toes must be used at times, but as little as possible.

Until recently, many debated methods and materials in ladder construction. The ingenious speleoinventor may enjoy reviewing the extensive literature and consider building his own. With diminishing use, however, the argument has largely subsided to a choice between two basic types: commercial cable ladders and simple homemade rope styles.

The commercial types are expensive, and their rungs have been known to slip at embarrassing times like National Speleological Society convention field trips. But they are generally quite durable. Rope ladders are cheap, flimsy, and designed to last no more than a few hard trips. Both are as light as possible. The cable ladder wins this round.

Rungs for rope ladders are easily made by drilling ⅜-inch holes in cheap softwood rungs just long enough and thick enough to handle a boot. The rungs are slipped onto ¼-inch or ⅜-inch manila or soft-lay or braided polypropylene or nylon rope and held in place by simple overhand knots twelve to sixteen inches apart. Luxurious models have an additional knot on the other side of each rung.

Such ladders are expected to break, and do. When jerked to a rude halt in midair, it is a remarkably foolish feeling to realize that you are instinctively, desperately clutching the useless end of the broken ladder. But this is part of the routine. The belayer lowers you, then the dangling ladder remnant. You tie the ladder back together and climb on out, vowing you'll burn that censored ladder as soon as you reach the top. Cable ladders break less often, which is fortunate, for they are less easily handled when they do break.

Surprisingly often, the ladder climber discovers that the belay rope has somehow snaked itself between rungs and through the ladder while on the way down to him. Usually this occurs disconcertingly high in the air, and going back down to start anew is no guarantee that it won't happen again. For this reason, every ladder climber should wear some form of harness. Then he can anchor himself to the ladder while untying his belay and using both hands to reposition it.

Ascent Knots and Devices

As rappelling devices have been the key to safe descent of deep caves, ascent knots and devices have been the key to safe return. The names of earlier types reflect their vogue among European mountaineers: for example, Prusik knots named for Dr. Karl Prusik, and Jumar ascenders manufactured in a tiny shop in Switzerland. The tremendous increase in vertical caving initiated by Bill Cuddington has triggered what is now a bewildering worldwide assortment of ascent knots, devices, and techniques. But one of three basic types is used for virtually all standing rope ascents in America: Prusik and similar ascent knots, Jumar and other hand-held mechanical ascenders, and Gibbs "rope-walking" ascenders.

PRUSIK AND OTHER ASCENT KNOTS

The Prusik knot is an age-old sliding hitch—a simple way of wrapping a sling rope around a standing rope in such a way that it slides freely up and down when unweighted and at right angles to the standing rope, yet engages firmly when angled by weight or pull.

To form a Prusik knot, a loop of smaller rope (a "sling" rope) is wrapped two or three times around a standing rope, each time passing through the bight, or eye, formed by the sling. A small space is left between each coil. The double wrap forms a four-coil Prusik; the

triple, a six-coil knot. The six-coil is increasingly being preferred to the four-coil—especially on muddy ropes, where the going gets tough.

Contrary to some assertions, this knot works equally well right- or left-handed ("upside down") on all ropes except laid types. On these it works best when the free end of the sling points toward the caver's right (reverse-lay ropes are virtually unknown on this continent).

In contrast to the helical knot discussed below, the Prusik is a forgiving knot, functioning well in most circumstances no matter what the type of rope or the conditions. If the sling rope has an especially soft surface like Dacron, however, the knots may be hard to loosen. When faced with extragooey Missouri cave mud or lava tube slime on the standing rope, two Prusik knots in series are said to be near foolproof. Mechanical ascenders may function better on ice or goo, but the Prusik does surprisingly well. Its cheapness and lack of damage to the standing rope make it especially popular.

Experts loosen and advance these knots with such speed and precision that onlookers often think that they need no loosening. This isn't quite true. Bob Thrun suggests grasping the sling rope with the third and fourth fingers, then lifting it and angling it like a pump handle, pivoted at the knot. This brings a little slack into the top of the knot and often does the job. If not, it may be necessary to squeeze the coils together or to push the doubled strand into the knot with the thumb and first finger.

THE HELICAL KNOT

The helical knot or pipe hitch functions similarly, but is constructed quite differently. Also known as the "ascender" knot and by several eponyms, including that of Larry Pemberthy, it, too, is a newly repopularized knot of venerable antiquity. It consists of several turns of a single ⅜-inch or ⁵⁄₁₆-inch sling rope round the standing rope. If nylon is used, six turns are usually employed, five if Dacron and four if Tenstron.

This pile of coils is anchored by a secure closure knot, usually a bowline. The long end of the sling runs downward from the top of the

92-foot entrance drop in Stamps Pit Cave, Tennessee, howing vertical ribbing and other features of domepits. Larry White is shown ascending by a standing rope echnique.

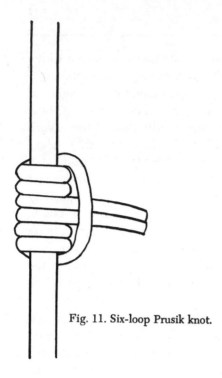

Fig. 11. Six-loop Prusik knot.

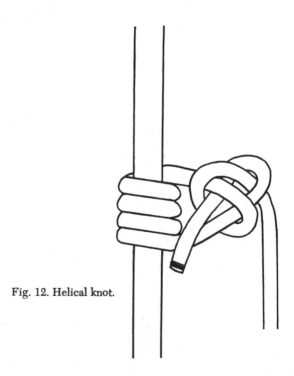

Fig. 12. Helical knot.

pile of loops (Bob Thrun recommends the bottom, but he's in the minority). When weight is applied, the pull angulates the standing rope and the pile of loops, compressing both. When weight free, and if grit is not excessive, it slides freely on the standing rope. Unlike the Prusik knot, it is easily tied at the free end of a fixed sling.

The helical knot is less durable than the Prusik knot, and the sling rope must be chosen for its compatibility with the standing rope (see Chapter 6), yet its enthusiasts insist that it never slips. While their number is comparatively few, one formerly held the world's record for a hundred-foot standing rope ascent with knots. On that climb (time: 1 minute, 12 seconds), Dick Mitchell of Seattle ascended so fast that his five-turn Sampson nylon knot partially melted and fused.

SEMIMECHANICAL KNOTS

Several sliding hitches include a carabiner for easier loosening. All work best on comparatively soft ropes. In the Bachmann knot, a sling loop is snapped in, then passed twice around the standing rope and the carabiner arm, then out through the carabiner. The carabiner is not intended to remain parallel to the standing rope, and normally rides at about a 45-degree angle unless pushed or angled to loosen the knot. It does not hold if inverted, and also tends to slide down the standing rope under its own weight. Beginners are apt to misuse the carabiner as a handle, but experts often find it a useful variant when Prusik knots stick annoyingly.

A somewhat different semimechanical knot is termed the RBS, pronounced "arbs." (The initials are those of Richard B. Schroeder, its popularizer.) It includes a sling loop snapped into a carabiner, then circling around the carabiner arm and standing rope. Since a foot or other sling need not be detached to use it, the RBS knot is especially helpful at ledges and other awkward spots. In use it distorts into many forms. It usually loosens readily when the carabiner is pushed upward at the same time that the sling rope is pushed upward into the carabiner. Its disadvantages are that it slides down ropes under its own weight and that it unscrews carabiner gates by its friction.

Fig. 13. Bachmann knot.

Fig. 14. Correct (*a*) and "upside-down" (*b*) Hedden knots.

a

b

OTHER ASCENT KNOTS

Several other types of sliding hitches are occasionally used as ascent knots. While none has gained wide popularity, investigations continue.

Mechanical Ascenders

GRASP-TYPE ASCENDERS

Vertical cavers may choose among several current mechanical ascenders. Except for the Gibbs (discussed on page 216), all are designed for hand operation. The Jumar ascender has been the overwhelming choice to date. All necessarily have both right- and left-handed forms. New models have a carabiner hole in a heavier frame than in the past. CMI produces a heavier ascender, much like the Jumar but with two carabiner holes at the bottom. These two ascenders are hand-sized "mechanical knots," shaped something like metal saw handles. Their use is shown in illustrations on pages 220, 222, and 223. Like all the ascending devices discussed in this book, they grip the standing rope with a toothed eccentric cog. Each has two controls: one (the "safety") opens to allow the ascender to be placed on or removed from the rope; the other (the "gate") controls the toothed cam, allowing the ascender to be slid upward. By "thumbing" this lever (holding it open with the thumb), it also functions as a descending device. Both controls lock in the safe position when released.

These ascenders are not foolproof. They sometimes slip on hard-lay ropes, and if a rope is not hanging free the safety may falter. Too, the cams have been known to freeze on icy ropes, and, even with care to prevent abrasion when moving them, they are a bit hard on ropes. With long use on muddy, gritty ropes, the teeth dull and the cams must be replaced. Often, however, reduced grip proves to be nothing more serious than grit between the teeth. Using a toothbrush five hundred feet down in the dark isn't as odd as it sounds. Every serious vertical caver should carry a safety Jumar-type ascender, no matter what his climbing system.

Some Jumars have been pulled off ropes accidentally, but with moderate care and practice they are reliable devices, not subject to the aggravating whims that sometimes befall knots. Starting at the bottom is especially easy. If the dangling rope drags upward, "thumbing" the cam of the lower ascender readily releases it. A safety sling, normally hooked to the seat from a foot Jumar, is a valuable fail-safe technique especially with the Texas system (see page 226). Those concerned with seat failure rather than accidental loss of a seat Jumar may prefer to attach it to a chest loop.

CMI markets another ascender in which the "handle" has been replaced with a single tie-in loop. Somewhat similar ascenders manu-factured by Clog (in Wales) and Petzl (in France) also are seen occasionally in this hemisphere. The basic Clog ascender has no safety latch, relying on a carabiner clipped through its lower opening to prevent accidents. Clog and Petzl also make so-called expedition models with heavy handles, but these are designed for mountaineering, not caving.

ROPE-WALKING ASCENDERS

Gibbs ascenders have cams that lock on the rope with comparatively little weight and release with slight upward motion. In contrast to hand-held knots and ascenders, these were designed for leg action and provide radically different methods of ascent. They must be assembled in place on the standing rope, and disassembled in place when the caver wishes to leave the rope. Yet they are extremely versatile and are used in innumerable ways never anticipated by their inventor and manufacturers. As a result, they now are used more than are Jumars. They are available in two forms: the free-running model (an improved version of an older "Quick Release Pin" model) and a spring-loaded model. The former is designed to slide freely along the rope unless weighted; the spring-loaded model stays in place on a standing rope and requires some effort for motion. In ordinary vertical caving, the free-running model is preferred overwhelmingly.

One Gibbs ascender is worn on the inner side of the lower foot, at ankle level. Originally the other was fastened to the opposite knee by an elastic strap or length of nylon webbing, with a safety strap attached to a body harness. The "shoulder" Gibbs and the "floating" Gibbs, however, are valuable new modifications. The floating Gibbs technique uses an ordinary foot sling for the second Gibbs, attached to a shoulder harness by a length of shock cord or other elastic material. This triggers the cam release on the floating Gibbs as soon as weight is released, pulling it upward. In racing, the floating Gibbs is usually worn at knee level; in caving, somewhat higher.

In caves, a third Gibbs is often worn at or just below shoulder level. If not, some other system must keep the caver upright and close to the standing rope should he lose consciousness. Some cavers use a wide ascender box. In the Cuddington system, a free-running Gibbs "safety" may be used above the top Jumar.

The Gibbs ascenders are particularly convenient in surmounting overhangs. They generally function well despite mud and wetness, even on icy ropes. Their chief disadvantage is their greater fixation (attaching oneself to the rope or changing ropes in midair just cannot be done quickly). Sometimes it is necessary to pull the standing rope under the opposite foot, or even to reach down and trigger the lower Gibbs by hand at the start of an ascent, but holding the free end of the rope tight against the boot soles usually suffices. They have been known to fail and their bindings to have come loose during ascents, but this is rare if properly rigged. If one fails or is lost, an "inchworm" ascent (see page 226) is feasible, though tiring.

When using Gibbs ascenders, the caver's center of gravity varies somewhat from that in other ascent methods. If packs cannot be hoisted separately, they should be hung from the legs, feet or hips rather than carried on the back. Some recommend the hips for all free climbs.

The type of elastic shock cord used to open the gate of the floating Gibbs is not standardized. Thick surgical tubing is widely used, despite its vulnerability to cuts. Straps designed for cartop carriers and other elastic materials are also used successfully. A spare is a necessity.

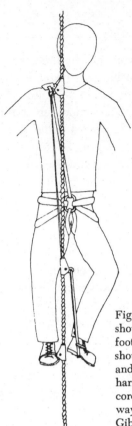

Fig. 15(*a*) shows caver rigged with shoulder Gibbs, floating Gibbs, and foot Gibbs (right foot). In this case, shoulder Gibbs is rigged with fore and aft straps. Other types of shoulder harness can be used. Elastic shock cord can also be anchored in other ways at both shoulder and floating Gibbs.

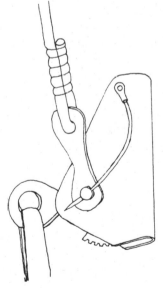

(*b*) shows attachment of the elastic shock cord to the floating Gibbs.

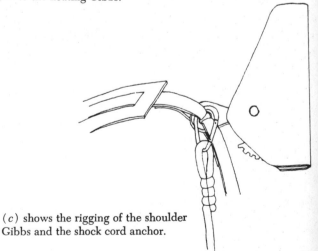

(*c*) shows the rigging of the shoulder Gibbs and the shock cord anchor.

MAINTENANCE

All mechanical ascenders must be examined frequently for excessive wear and lubricated periodically, and before each use their slings should be checked for abrasion, fraying, and cuts. Carabiners similarly require frequent inspection for corrosion of hinges, pivots, and other hidden areas, then lightly oiled. Loren Bollinger recommends silicone spray for rust prevention. He also suggests agitated soaking in hot water with a little detergent and household ammonia as an easy way to get rid of grit and sand.

THE ASCENDER BOX

The first time Keith Wilson tried his new invention—the ascender box—in a truly deep pit, he climbed 1,040 feet in thirty-three minutes. Only the gasoline-powered Mechanical Ascending Device beat him out of Sotano de las Golondrinas.

The box, a very efficient user of many ascent methods, consists of a simple housing attached to a chest harness, which keeps the caver close to the standing rope. It contains two parallel pulleys, which direct the rope rather than provide mechanical advantage, one for the standing rope, the other for a sling if required by the technique chosen. Two long bolts loosen enough to permit the rope or sling to be inserted or removed without tension. A carefully machined model—the Gossett ascender block—is increasingly popular, although dangerously sharp edges on some units need filing or buffing before use.

With the ascender box, chest harnesses are worn extra-snug and high. The Gossett block also must be snug against the body but may be worn as low as hip level. For both, the main strap commonly is 2-inch seat-belt webbing. Carabiners can be used instead of the box, or the Mitchell system (see page 225), but at the cost of serious energy drain.

Bill Cuddington (*at right*) coaches
Russell Gurnee on the use of the
ascender box and Jumars in backyard
practice.

In free-drop ascents, these devices assist some cavers to switch
hands in midclimb when one arm becomes overtired. Some experts
lose hardly a stroke. New pit cavers do well to practice this from the
start. Some old hands have developed such fixed patterns of coordination between hand and foot that they find they cannot switch without
disrupting their entire rhythm.

Ascent Systems

Ascent knots and devices are used with foot slings and body
harnesses in more than a score of "systems." Each originally had
enthusiastic partisans. With increasing knowledge and practice, how-

ever, vertical cavers are increasingly recognizing the specific values of various systems and applying them to specific needs, switching from one system to another as circumstances require.

THE CUDDINGTON SYSTEM

Bill Cuddington has clarified these matters by pointing out three phases of ascent: (1) free-drop pitches, or others in which the wall does not interfere; (2) coming over the lip at the top; (3) wall walking. The key to the Cuddington system is his carrying an extra (third) Jumar ascender in his pocket, already hooked to the carabiner of his Swiss seat. This provides great flexibility in shifting from phase to phase.

In the primary phase, Bill employs the ascender box, with long and short foot loops attached to Jumar or Bluewater ascenders. This combination is well suited to most of the deep pits sought out by advanced vertical cavers.

The most critical point in standing rope ascents is usually at the lip, where the climber's weight pulls the standing rope tight against the rock. Here the prepared climber can hop nimbly into phase two of the Cuddington system by clipping the third ("safety") Jumar to the standing rope just above the box. He then "comes out of the box" (vertical cavers' cant for removing the ropes from the box) and goes into the so-called Texas system: use of his short foot sling plus the seat Jumar. The long foot sling is out of the way until it can be used advantageously above the lip. This also works well when free pitches up to thirty or forty feet high are encountered during wall walking.

On steep, slanting walls, the ascender box pulls its users off balance, into the wall, and thus cannot be used effectively. For phase three Cuddington merely hooks his third Jumar onto the sling of the long foot Jumar (a modification of an early two-knot system mentioned below). On long ascents he reduces leg and arm strain by ensuring that the legs are accomplishing equal work. To do this he shortens one of the foot slings by wrapping it around the arm of a carabiner. He reports excellent speeds on near-vertical slopes as high as nine hundred feet.

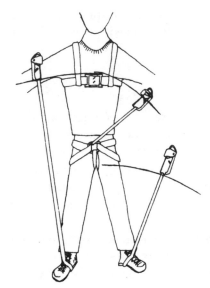

Fig. 16. Rigging for the Cuddington system. Shown are short and long foot slings with Jumar ascenders, ascender box and harness, and third "safety" Jumar attached to Swiss seat and kept in a pocket during the Cuddington primary phase.

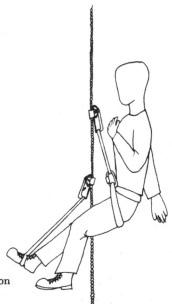

Fig. 17. Phase II of the Cuddington system (the Texas system).

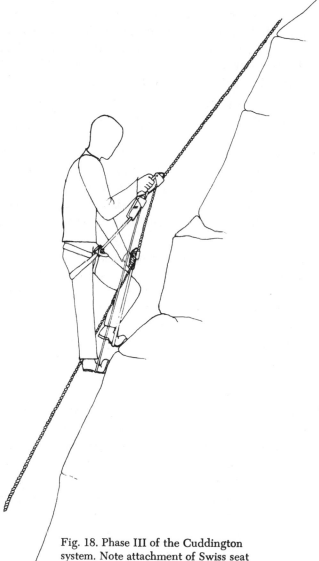

Fig. 18. Phase III of the Cuddington
system. Note attachment of Swiss seat
Jumar to long foot sling.

THE PLUMMER AND OTHER THREE-KNOT SYSTEMS

Several versions of the versatile three-knot system are in common use. Pioneer Bill Plummer's "sit-on-the-heels" methods functions well up to about two hundred feet, and is quite popular in parts of the eastern United States. Its first step is adjustment of the chest harness and its sling so that its knot is at forehead level when some of the climber's weight is placed on it. The foot slings are about three feet long, and are attached to the rope where the hands fall naturally upon them.

The climber then squats on his heels, his knees a few inches off the ground. He adjusts his foot slings so that their knots are just below the upper knot, roughly about three feet above foot level. The knots must be separated by at least a finger's breadth. The upper foot sling is two to three inches longer than the other, so both feet can be brought to the same level.

The caver then stands erect, keeping the standing rope close to his body. Quickly he grasps it with one hand, as high as possible, and slides the chest knot upward a foot or more with the other hand. As his weight goes onto the chest sling, the foot with the longer sling must be lifted for balance. With all weight off that foot, its knot is slid upward until the heel is close enough to sit on. The climber immediately shifts his weight onto that foot to get it off his chest. Then the third knot is slid upward and the caver rests his weight on both feet in a squatting position, his center of balance as close as possible to the standing rope. Then he stands up, slides the chest knot upward, and continues.

If the dangling rope sticks and comes up with the lower foot, this is usually controlled by scooping up the rope with a foot, grasping it at waist level, and straightening and tightening it with the other foot. It's easier if someone else pulls on it.

This "sit-on-the-heels" system substitutes the sitting position for a Swiss seat, but some have found that connecting the chest harness to a Swiss seat relieves part of the strain on the rib cage. Some also find that shorter foot slings are more to their liking, or one long and one short foot sling. The latter modification is regularly used in the southeastern United States for five-hundred-foot free-fall ascents, and has served uneventfully for twice that distance.

When approaching the lip with a three-knot system, temptation is great to remove the chest knot and replace it above the obstacle. Several trying this have suddenly found themselves hanging upside down by their foot loops, which is the hard way to test them. If the lip cannot be surmounted without removing that knot and placing it topside, the upper foot loop should be converted into a temporary second chest loop, or the Texas system used.

THE MITCHELL SYSTEM

Economy of motion is obviously desirable when compatible with safety. Many vertical cavers have tried systems in which two rather than three knots are slid up the rope (eventually such systems as the Cuddington phase three evolved). Those who attempted to abandon the chest or seat sling rather than hook it onto the long foot sling intermittently found themselves in the unenviable upside-down position just mentioned. Dick Mitchell found a cure for this, running his long foot sling beneath a chest loop or harness. Although now largely replaced by the ascender box, it still may be useful in emergencies.

THE PYGMY SYSTEM

The Pygmy system is named for its two short foot slings, but it also uses a shoulder Gibbs and Cuddington's spare Jumar. By wearing proper harnesses and also an ascender box, exceptional flexibility can be achieved.

TWO-KNOT SYSTEMS

For a lengthy free-drop ascent using knots, some two-knot systems with foot slings of different lengths are probably the most efficient yet devised. While some prefer the reversed rig used in Cuddington's third phase, the Plummer version attaches the upper foot sling to the chest sling just below the ascent knot or device. It requires particular precision in sling lengths and the tightness of the chest sling. Originally it included foot slings of virtually equal length but this causes undue strain after about 150 feet. It is especially useful on short to medium free ascents, and against not-quite-vertical walls up to about two hundred feet high. Although not as fast as the Cuddington Jumar system, similar two-knot systems are satisfactory for pitches of several hundred feet and function well against walls.

THE TEXAS SYSTEM

The Texas system has already been mentioned. It somewhat resembles a three-knot system with the chest sling moved down to the Swiss seat and one foot hanging free. It can be used in free-drop ascents when an injured leg is out of action but is especially useful when some wall walking is necessary and when coming over the lip. A safety string should link the foot sling to the locking carabiner of the Swiss seat.

THE "INCHWORM" AND ITS VARIANTS

A funny-looking, rather divergent system was quickly dubbed the "inchworm." This moderately efficient technique apparently began when Charles Townsend fitted a Jumar directly onto a bar for both feet. The caver hangs briefly by a chest or seat Jumar or ascent knot while he bends his knees, raising the feet and what some call the "mar bar." Then he stands erect while he raises his chest knot, and so on.

Many modifications exist. Peter Strickland has used a refined version successfully in the 1,040-foot ascent of Sotano de las Golondrinas. Others prefer two foot loops attached to a single knot or Jumar, at about knee level (this is also known as the Number One Texas System). A wrist loop is helpful in getting started, and a short Jumar

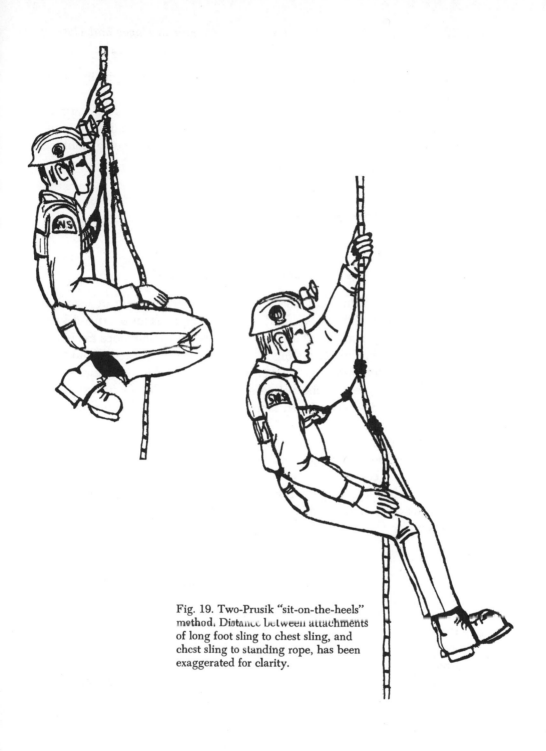

Fig. 19. Two-Prusik "sit-on-the-heels" method. Distance between attachments of long foot sling to chest sling, and chest sling to standing rope, has been exaggerated for clarity.

chest sling eases arm strain by opening the cam automatically. This, however, causes rope abrasion.

The "inchworm" is forgiving, and easily learned. Thus its variants are often successfully used by beginners and others faced with comparatively short ascents. The commonest problems are being hobbled in wall walking and chimneying.

RANDOM NOTES ON THE ASCENT SYSTEMS

FOOT LOOPS AND "CHICKEN LOOPS"

In all these techniques, the foot loops should be snug against the midsole. Many methods are used, some as simple as an overhand knot in a sling. Particularly neat are sewn, prefitted webbing loops anchored to the lower end of the sling by a "half Prusik." Each should be secured by some form of "chicken loop" holding the foot sling to the ankle when the seat or chest sling goes and you suddenly find yourself swinging upside-down, dazed from thwocking your head on the wall during the swing. This is so valuable a conservation technique, effectively avoiding littering caves with blood and guts, that nobody seems afraid of being called "chicken" any more. A short, separate loop can be used, or a continuation of the foot sling. As already mentioned, the Prusik auto-belay, or free-riding second sling hooked to the seat or chest harness, avoids much of the inconvenience inherent in this position.

ASCENT CYCLE LENGTHS

In all ascent methods, momentum is ideally channeled in a smoothly coordinated, rhythmical progression. Many cavers routinely gain more than twelve inches with each cycle, although for the first few feet eight-inch cycles may be easier. Once the rhythm is flowing, many can maintain a twenty-four-inch "bite." Some aim to gain three feet with each cycle, but for all but the lankiest this requires excessively fatiguing arm action.

ASCENT SPEEDS

Underground, racing speeds are rarely appropriate. Many consider five to fifteen minutes "about right" for a pleasant hundred-foot ascent. Faster ascents are obviously tiring. Less obvious but similar energy drains result from overlong stays on the rope. The three-knot system is especially popular because it produces near-automatic pacing as its cycle is repeated over and endlessly over. In skilled hands the pace causes no greater strain on the rope and anchor than 20 percent of the caver's weight.

UNDOING SNARLS

Especially on laid ropes, the three-knot system is somewhat prone to spin, which jumbles the slings. Spinning the reverse way rarely helps. Each knot must be loosened individually and slid around the rope to its proper position.

CHANGING ROPES

With the Cuddington and several other systems, changing in midair from one standing rope to another is no particular problem. The seat Jumar goes onto the second rope in the Cuddington system, the caver "backs down" on the original rope until he settles comfortably onto the second, then comes out of the box and moves it to the second rope along with the long foot sling. The short foot sling follows.

Tandem Climbing

When a standing rope is long and a party large, two can climb reasonably safely on a single rope if they stay less than eight to ten feet apart. Perhaps three on a rope, on each other's heels, but two is really safer. Because of the added weight, ascent knots and devices react unexpectedly, and erratic motions trouble all concerned. Except possibly for inchworm tandeming in which nausea is a particular problem, however, the limiting factor is falling material: rocks from the lip or ledges, pocketknives and their blades, carabiners, rappel racks, cameras, flashlights, carbide lamps, canteens, cans of Sterno, beans, fruit, or carbide—the damnedest things whiz-z-z-z down pits below climbers. An expert pit woman recently laughed at Bill Cuddington's concern, started up ahead of her rope mate and promptly dropped a Jumar on his head. Fortunately the Jumar fell only ten feet. At twice that distance and four times the speed, it would have penetrated more than hardhat.

The upper climber may have trouble with slipping knots because of the weight below. Normally, the slowest climber should ascend first unless he has knots and the other climber has Jumars. Some snicker that the last man on the rope should keep his carbide flame extra long to spur the next to ever-greater vigor. The Cuddington system permits cavers to change places on the rope if necessary, or a climber to shift to a short second standing rope at the lip.

Practicing in the Dark

Some vertical cavers routinely climb without light. Others have done so, accidentally. Practicing in darkness is a good idea. Especially for carbide cavers who keep singeing the rope.

Lending Equipment

Except in instruction and experimentation, lending vertical caving gear is a virtual invitation for an inexperienced caver to get in over his depth. Anyone serious enough about vertical caving to try these single-rope techniques will want to develop a rig in which he feels comfortable. And practice, practice, practice. Somebody said it correctly a long time ago: neither a borrower nor a lender be.

The Mechanical Ascending Device

The Mechanical Ascending Device (affectionately or otherwise known as MAD) is a gasoline-powered brainchild of the Nittany

Grotto of the National Speleological Society. Weighing about twenty pounds, it has accomplished the 1,040-foot entrance drop of Sotano de las Golondrinas, miles deep in roadless Mexican mountains. There it was set for a speed of thirty-three feet per minute, less than half its theoretical maximum. Its fuel consumption averages two miles per gallon, or two to three thousand feet per tankload (one quart). Although rugged and apparently reliable, its future appears clouded because of its noise and atmospheric pollutions.

Other Ascending Devices

SCALING POLES

Where feasible, much time can be saved by hoisting rope or cable ladders into position with scaling poles. These are tricky and dangerous, however, and probably should never be used for ascents of more than thirty feet. Most consist of sections of light alloy or steel tubes, which screw or bolt together. Their screw threads and other connections are easily damaged. Metal fatigue and internal rusting cause

Scaling poles are tricky and dangerous, but sometimes necessary in first ascents underground.

other problems, so each must be carefully tested on the surface before each use. The pole must also be as nearly vertical as possible, and, unless the ladder can be hooked over an unusually secure projection, guy ropes or cables are usually needed to steady the pole.

Belaying the first climber is usually difficult. Pitons fifteen and twenty-five feet above the floor may suffice for a thirty-foot ascent. Sometimes a rope loop for a pulley belay can be hooked over a secure nearby projection before the pole and ladder are fixed in position. Obviously, such a belay must be tested with hard practice falls.

TYROLEAN TRAVERSE

The rule on underground Tyrolean traverses is: Don't try them. It is far better to allow a rope or ladder to hang free in a void, then descend one side and ascend the other. Sometimes rescuers may need to transport a victim on a tight double line (see Chapter 10), but even that can usually be avoided.

One of my own not-so-bright ideas was to convert the side of a pit into a chimney by a variation of this technique. After three or four tosses we wedged a rope ladder in a crack on the other side of the pit and tightened it nicely. On belay, I began what is now fondly recalled as the Northwest's Famous Foot-Fanny Traverse—my back to the wall, feet inching along the straining ladder. To reduce any pendulum effect, I worked a sling rope along the ladder, too. It worked, but just barely. Definitely not worth repeating.

Once across such a pit by descending and ascending, the return is simple. A rappel rope is rigged to a descending ring, and the last caver rappels to the bottom, pulls that rope down, and climbs out on the other.

The Future

The speed of evolution of these techniques is amazing, near incredible. The world of North American caving can almost date itself B.C. or A.C.—Before Cuddington and After Cuddington. Only now are we at about A.C. 30.

Yet today's vertical triumphs are but prologue. A thoughtful midwestern caver recently mused upon his trip to Alabama: ". . . kids between the ages of three and thirteen play on ropes like we in Missouri walk through mud. There are kids in Huntsville that will have more vertical experience before they leave grade school than most Missouri cavers will get in a lifetime . . ." Nor was he exaggerating. At four and a half, young Scott Stokes routinely Jumared nearly a hundred feet without even working up a sweat. What the coming generation will achieve is beyond prediction. Maybe it will even learn how to climb and use pitons.

Or maybe its members will grow up shorter. Bill Cuddington relates a previously unreported incident that may hint an ominous trend. It seems that in Georgia's Ellison's Cave, a hapless caver one day misused his rack. When he left the hospital, he found that he had lost an inch of height.

Bill is resigned to finding others "up to their hips in solid limestone." There's much more to vertical caving than fun and games.

Maybe this chapter will help.

9

Cave Medicine and First Aid

Blinding light fills the pit, and sound beyond belief.

Seventy feet up a swaying rope, an unconscious caver slumps on his Jumars, motionless, pulseless. For a long second, his dazed fellows gape, then scramble desperately for their own ascenders. As long expected, a lightning bolt has finally struck a pit plunger. Through anticipation and practice, telephone and electrical linemen have saved the lives of several of their fellow crewmen seemingly electrocuted in a similar situation. The cavers have a little more than three minutes to find out if they can do as well.

As aboveground, the watchword of spelean first aid is prevention: don't become a statistic. Yet this is not wholly possible for novice or expert. The caver must know a specialized form of first aid as he knows his belay, his compass, his headlamp.

The principles of first aid in caves are those of the surface, but their successful underground application may differ enormously. In some areas cavers teach other cavers first aid in a highly effective form scarcely recognizable by the American Red Cross. Elsewhere, any competently taught course can serve well if combined with intimate familiarity with the cave environment. Mountaineering or industrial versions are even better, and the U.S. Department of Transportation's superb Emergency Medical Technician course is now increasingly available. Inexpensive Red Cross and other excellent manuals are so widely available that spelean applications and not the techniques themselves rate the space here.

The Impact of Injury

When a major accident occurs underground, every member of the party immediately suffers serious emotional shock. The mere knowledge

High on the Wyoming side of the Teton Mountains, the yawning mouth of Wind Cave reveals no clues to the cause of the fatal lightning bolt that struck here some years ago.

that this is both expected and readily controllable, however, is usually sufficient. Even novice cavers characteristically brace themselves and, calmly or jittering, get on with the job of determining what must be done, and how soon.

Balancing the Risks

Aside from rare life-endangering situations like cardiac arrest, obstruction to breathing, perhaps severe arterial bleeding, and the like, the first need is to balance the severity of the injury against the inherent hazards of the situation. If the risk of hypothermia is low, a badly injured caver is often best treated where he lies, even for many hours. Some years ago, Carroll Slemaker suffered a compound leg fracture some 330 feet down in Church Cave, California's deepest. His fellow cavers kept him warm, in good spirits, and fairly comfortable overnight. Meanwhile, a full-scale rescue operation was organized and his fellow cavers were able to pass their responsibilities to a fresh team. A noncaver physician was belayed up a long, steep gulch to the cave, thence to Carroll. After injection of an opiate, the victim was able to assist his rescuers enormously during the slow, difficult exit.

The other extreme is obvious. A wet, chilled person with a minor injury must be warmed immediately, then evacuated as rapidly as is safe. The dangers of hypothermia have already been mentioned more times than the Texas caver may consider necessary. Yet, merely lying strapped in a Stokes stretcher outside the narrow inner entrance of Idaho's Papoose Cave during a practice rescue, Clarence Hronek—father of Vancouver Island caving—came near succumbing despite seemingly adequate blanketing. Waterfall spray, active air currents, and fatigue have caused several near-fatalities in that chilly cave; I have been affected myself. Nevertheless, application of antihypothermia principles by experienced cavers led to rapid success in the single rescue necessary there to date.

Cardiac Arrest

The electric discharge of lightning extends surprising distances into caves. Some years ago a group of hikers sought shelter in the mouth of Wind Cave high in the Teton Mountains of Wyoming (not the famous South Dakota cave of the same name). When lightning struck, one died.

The same fate might be caused by faulty wiring in commercial caves and possibly by faulty strobe flashguns. And while closed chest cardiac massage will often restart the heart in such cases, and in some cases of apparent drowning, it can also kill. Even under more favorable circumstances than most caves present, this treatment often breaks ribs, which is no great problem. But it can also push them into the heart or lungs.

Such a catastrophe rarely occurs when the approved technique is followed calmly, accurately, and under good conditions. Before starting, however, everyone at hand must be absolutely sure that no heartbeat is present. This can be checked in several places—at the apex of the heart, for one, by placing the palm against the middle of the left rib cage. Other checkpoints are in the neck and in the notch at the top of the abdomen, where the ribs arch together. The wrist pulse is likely to be misleading at such a time; many a rescuer has hopefully counted his own pulse, thinking it was that of the victim (this is easily checked by counting the pulse of the person counting the victim's pulse). The femoral pulse in the groin is stronger and more reliable, but sometimes hard to find.

No one may delay while dithering, checking and rechecking. After more than three minutes some stopped hearts can be restarted, but the brain is dead forever.

To pump blood through the heart and lungs in this way, the sternum (breastbone) must be depressed about two inches with each stroke. This is easy on an unconscious person—too easy, in fact—but is much more than can be done on a conscious person, tensed up to make sure that his partner isn't going too far with *his* heart. Practice dummies ("Resusci'Annes") are widely available in the United States and Canada. Every caver should practice with one.

HEART MASSAGE ON A ROPE

More than one caver has been knocked off a ladder by lightning—fortunately without serious effects, as far as I have been able to learn. Sooner or later, however, some pit plunger will be electrocuted while on a standing rope.

Ordinary resuscitation training stresses the need for a firm, flat surface. When a caver is pulseless on a rope, anything resembling such a surface is manna from heaven. Yet raising or lowering him is likely to require much more than the three minutes available before the brain dies.

But the victim need not be counted dead if his mates are prepared to begin effective resuscitation on the rope. Especially if the chief rescuer is endowed with powerful arms and shoulders, and has practiced a time or two, this is much less difficult than it might seem.

The climber's slings must be shifted so that he hangs somewhat head down or sideways. The rescuer positions himself behind and close under the victim, pulls him into a sitting position, and, if necessary, anchors the victim to him. The victim's pulse, breathing, and mouth are checked. Chewing gum, dental plates and/or loose teeth, dirt, guano, vomitus, or other obstructions must be cleared away instantly and effectively.

The rescuer first swings the victim so that he can breathe five times mouth to mouth or mouth to nose. This can be done at the same time as the checking for pulses, especially the carotid pulse in the neck.

Next he half spins the victim and thrusts his arms around the limp chest, grasping his own wrist, cocked "so that hand is like a plunger on the lower half of the victim's sternum," to quote *Northwest Medicine*. He gives about fifteen two-inch compressions, then spins the victim for five more breaths, and so on. And on and on and on.

FAILURE OF RESUSCITATION

The tests for effectiveness of closed chest cardiac massage are simple. If the victim regains consciousness within a few dozen strokes, the technique was right. If he does not, the artificial heartbeat is still effective if a pulse can be counted at the wrist.

The decision to give up is never easy, and there is little danger of quitting too soon. Skilled resuscitation can maintain an effective pulse and respiration for hours. The chance of the heart resuming its normal beat, however, becomes less and less as time drags on and on. If a strong, palpable pulse cannot be maintained, the brain and other vital organs will suffer permanent damage in a comparatively short time. No specific times can be given because the situation is so variable.

The eyes can give you an idea of success or failure. On the surface, huge, fixed pupils that do not recover with effective resuscitation are usually accepted as a sign of failure. In the darkness of a cave, this is much less valid. If the pupils become smaller with the flick of a bright light, hooray! If not, keep on trying for a while longer.

If things were going well and the risk of hypothermia was low, I would like my own rescuers to try for about an hour. If I were on the other end, I'd expect to try twice that period, although the last hour probably would be nothing but hope, hope, hope.

If the passing of time is an increasing risk to others, much less than an hour should be spent. Most successes come in the first few minutes.

ARTIFICIAL RESPIRATION

If the victim is not breathing but has even a weak pulse, the situation is far happier. A few seconds must be devoted to clearing the mouth and any external pressure on the windpipe—chinstrap, rope, or whatever. Then mouth-to-mouth or mouth-to-nose breathing can be applied in almost any position or location. The rescuer must expect vomiting, for the stomach balloons and empties automatically, again and again. Virtually every caver will tolerate this gladly, however, knowing that the victim would do at least as much for him. A face-down position is obviously desirable (so that the vomitus escapes freely and not into the windpipe), but a compromise position must often be accepted as the best available.

If possible, this artificial breathing should be begun without moving the victim (presuming he is not face down under a waterfall or in equally desperate straits). First, a few dozen breaths into the unconscious caver, then the situation can be studied. If the apnea (cessation of breathing) follows a fall, there is probably a severe head and/or neck injury. It is essential to avoid making a bad situation much worse through unwise moves. If the problem is due to near drowning or electricity, the victim can be moved to the most favorable position available and he will usually respond in a few minutes. If not, the artificial respiration should continue as long as there is a pulse—and a few minutes more. Even after the heart stops, there is still a slight hope through a short trial of closed chest cardiac massage.

Chest Injuries

If something jagged has dug a hole into someone's chest so that air sucks in and out, that hole must be made airtight quickly or he will probably die within a few minutes. While it is obviously nice to avoid getting dirt, germs, cactus spines, and other contaminants inside the chest, this is of far less importance.

Much air is trapped when such a hole is sealed with a wet handkerchief, plastic bag, or whatever. Once it is airtight, it may be possible to plop the seal off and on, to allow some of this trapped air to escape when he breathes *out*, without allowing it to be sucked back in when he breathes in. This would be a great boon to his laboring lungs and heart, but is difficult to accomplish. Getting the sucking wound airtight is much more important. It's better to seal the hole when he has just breathed *out*, but such victims breathe too rapidly for this to be much more than theory.

The victim with a considerably crushed chest faces a somewhat similar problem. One or both sides of his chest wall suck in when he tries to inhale, pouch out when he tries to breathe out. If severe, the worst side must be stabilized within minutes to prevent death. If all else fails, his arm—bent at the elbow—can be wrapped tightly to his chest. A wide elastic or other bandage holding a pile of padding tightly to the chest is usually better. Such padding will push the flailing part of the chest inward and reduce the victim's breathing space, but this is usually better than the ineffective flail action. Just don't overdo it.

Bleeding

Bleeding is rarely if ever immediately fatal underground, but a large artery might be cut by a sharp rock or piton. This would be a grave emergency, necessitating rapid control of the bleeding (as taught in first-aid courses) before the victim bleeds to death.

The large artery on the inner side of the upper arm is about the only exposed location that can be managed with a tourniquet and pad (must I add, never put a tourniquet around anybody's neck?). Otherwise, tourniquets are likely to do much more harm than good in caves. If one must be used, someone must find a way to write the time on the victim's forehead. Otherwise the rescue team will not know when it is time for the short relaxations of the tourniquet that may help prevent gangrene.

Lesser bleeding is lower on the priority list, even if seemingly very severe. Usually direct pressure applied through a clean handkerchief is sufficient. Don't worry about scalp cuts. Almost always they bleed alarmingly, then stop spontaneously. If the victim is unconscious, they are best left untouched. A jaggedly fractured skull may be hidden under the bleeding point, and bone fragments just don't mix well with brain.

Shock

Shock is treated as recommended in first-aid manuals, with the addition of precautions against hypothermia, which worsens the problem.

Head and Spine Injuries

After life-threatening emergencies are under control, the most critical first-aid skill in caves is the handling of neck and back injuries. The basic principle is to do no further harm. Every year, mishandling of accident victims in broad daylight, in comparatively easy locations, causes tragic, needless paralysis. That grave hazard must be balanced against the risk of death from hypothermia and shock. Severe pain in the back or neck does not necessarily mean a severe spinal injury or a risk of paralysis from incautious treatment, but sometimes it does.

Three situations require special consideration: (1) the unconscious person, (2) the conscious person with what seems to be a severe back or neck injury, (3) the conscious person with what seems to be a minor injury but pain in the back or neck.

SPINAL INJURY IN THE UNCONSCIOUS PERSON

Assuming that the victim is reached quickly, virtually everyone unconscious after a fall or blow from a falling rock has suffered a serious head injury. And, caves being what they are, major head injuries are almost always associated with additional neck or back injury.

No one can be sure how badly such a patient is hurt unless he awakens and says that he has no pain along the spine, nor paralysis. No shooting pains, nor sensations as if electric shocks were coursing down his legs or arms.

Thus, every unconscious patient must be treated as if he has severe combined injuries.

If the victim is under a waterfall or in a pool or stream, obviously it may be necessary to move him immediately to prevent his drowning or within a few moments to prevent hypothermia; even an intact wet or dry suit will protect an unconscious person for only a few minutes.

But each unconscious victim must be moved with his neck and back treated as a fragile crystal—perhaps a gypsum needle six feet long and a quarter inch thick. Most fractured necks will tolerate a little *backward* bending, the lower spine a little more so. Neither will tolerate even a little forward bending, or bending to the original position *after* a little backward bending.

Once the victim is in a warmer, drier place, it may well be worth waiting a few minutes to see if he will awaken and say where he hurts. Or admit he's one of the foolish minority of epileptics who try to ignore and conceal their problem until it becomes everybody's.

But suppose he doesn't awaken?

If the legs are rigid or convulsing, or if the pupil of one eye is larger than the other, or if he is bleeding deep in the nose or ear, he probably won't awaken for hours or days. (Occasionally, however, people are fooled by a minor nosebleed or blood that has run in from the outside.) Such patients present perhaps the ultimate problem in cave rescue. Were the victim in a hospital bed instead of the cave, his chance for life in even a crippled state is small; hypothermia may be less dreadful than some of the alternatives. Unless death appears virtually certain, such a victim is often best treated by much the same techniques used successfully in the Slemaker rescue mentioned earlier in this chapter.

THE CONSCIOUS PATIENT WITH SPINAL INJURY

If the victim is able to talk to his rescuers, the grave uncertainty is greatly reduced. Severity of neck or back pain usually parallels the risk of paralysis from mishandling, but the location of the pain is important. Pain in the very low back carries comparatively little risk. However, with pain in the lower dorsal spine (the part the ribs hook onto—and the commonest danger area after a fall in the upright position) mishandling carries a high risk of paralysis of the legs.

Neck pain is somewhat less reliable. Some broken necks with grave risk of paralysis below the shoulders are surprisingly pain free.

The risk is extreme if the victim has the feelings of electric shocks mentioned above, especially if they are in the arms as well as the legs. The presence of partial paralysis is the worst of all. Such paralysis is occasionally temporary and merely a response to the shock of the fall. If due to spinal injury, however, any except the most cautious movements are likely to cause complete paralysis. On the other hand, if the victim can move his neck and back despite pain, and has none of these special danger signs, the risk is comparatively low, and he may be able to get up in a few minutes and go about his own rescue. It is best to err on the side of caution, for some patients with broken necks have walked into hospital emergency rooms. A few never quite made it.

Fractures and Sprains

In comparison with the problems of head and spine injuries, those of arm and leg fractures seem easy and pleasant. As aboveground, "splint 'em where they lie" is the principle.

Few ordinary splint materials grow underground, so ingenuity reigns. Often the best splint is a fellow caver, or the patient's own body. Arm fractures may be bound to the chest. Leg fractures often ride well if immobilized to the other leg. After Carroll Slemaker had been given an opiate, it was possible to remove his splint and substitute traction by a rescuer while passing through a crawlway. At the far end, however, his foot was found to have turned 90 degrees.

Plaster and other formal splints are usually helpful only if evacuation may be delayed safely. Rolled wire mesh splints are perhaps the largest that can be carried in the normal cave pack. Pneumatic types that immobilize a limb or an entire patient are very useful if the cave is large enough. Otherwise they puncture. And, as first-aid courses teach and their students forget, boards and similar hard material must be well padded to prevent serious nerve and other damage.

Compound fractures are not as grave as in the preantibiotic era. Slemaker's leg fracture, for example, was compound and his evacuation delayed. Yet the outcome was excellent. Nevertheless, it is common sense to prevent broken bone ends from jabbing through the skin where crippling infection lurks. If this has already occurred, a clean handkerchief is the next best thing to a sterile dressing.

Minor fractures differ little from severe sprains, and can be treated as such. If an injured caver can use an extremity despite the pain of a severe sprain or minor fracture, the additional damage caused by its use is often minor, especially in comparison to that resulting from delaying rescue so that he need not use it. If someone has higher boots than the victim of an ankle sprain, trading a boot and lacing it high and tight may suffice. Wrapping a sprained knee tightly, then walking and crawling and Prusiking, one legged or stiff legged, is often better than trying to splint it.

Shoulder Dislocations

Some first-aid manuals include detailed instructions for replacing dislocated shoulders. In cave accidents, it is usually best not to try this unless the shoulder has been dislocated many times before, so that it pops in and out without much stress. In this case the owner can usually tell others exactly how to do it. Otherwise, the arm should be bound to the chest and the patient evacuated.

Burns

Caves might seem the world's most unlikely places for severe burns, but they do occur. Chemical burns from leakage of wet cells are covered in Chapter 4. If no neutralizer is at hand, copious water is the remedy. For thermal burns, the best treatment is cold water if the blisters are unbroken. Broken blisters and charred third-degree burns provide channels for severe infections, so cave water should be avoided. If any treatment is needed beyond exposure to cool air, it should be limited to canteen or boiled water, not water purified by chemicals. A little water can be stretched a long way by placing two layers of moistened cloth (preferably clean handkerchiefs) over the burn. The outer cloth can be removed periodically, resoaked, and waved vigorously to cool it.

Hypothermia and Heat Exhaustion

As already stressed, hypothermia is the chilling of the vital organs of the body's core to the point of decreased function or death. Heat exhaustion is the reverse. In addition to the preventive approaches indicated in Chapter 5, rapid warming of the heart is the cure for hypothermia. Direct heating by hot liquids taken internally is best, preferably by mouth. The liquid should be tested in other mouths before making the victim swallow it. Ideally, it should be so hot everyone can barely stand it.

Next best is external heat, especially at points where it is rapidly transmitted to the heart—the abdominal wall, neck, and chest. Carbide lamps and candles make good heat sources over which to huddle. In an emergency the abdominal and other skin surface of the warmest available caver is more likely to be lifesaving. Breathing heated air also is effective, but it must not contain deadly gases released by open flames and catalytic converters.

Frostbite

Frostbite is rarely a problem in North American caves because most large freezing caverns are in locations where proper precautions must be taken outside as well as in. A few glacier caves are at surprisingly low elevations, however. Here one's feet may have become soaked wading a creek or slogging through slushy snow without proper preparation. When feet stop feeling cold when they should be feeling colder, *beware!*

Because of the difficulty in judging the correct water temperature, body heat is usually better treatment than warm water. The ideal form of body heat is the bare abdomen of whoever talked you into this particular mess. If it's your own darn fault—well, it helps if you're double jointed.

Immersion Foot (Trench Foot)

When the feet remain wet and chilly for considerable periods, the blood vessels undergo permanent damage. The effect varies from person to person, but is cumulative: after a caver has had a mild case, next time he'll have more of a problem, and sooner. The early symptoms are somewhat like those of frostbite: the feet are cold, painful, and somewhat numb, conditions that last for a time even after returning to a warm, dry area. Precautions include use of wool socks, powdered talcum (no, it doesn't have to be perfumed baby powder!), and careful attention to drying and warming the feet. While predominantly a problem of cold water, this can also develop in much warmer climates.

Animal Stings and Bites

Gleeful cavers relate dreadful yarns about past, present, and potential misadventures with both ends of an enormous menagerie. Most are nothing but tall tales. Except for the pincers of cave crayfish, which are hardly much of a threat, the blind fish, salamanders, and other troglobites of North America can't even defend themselves.

Many large animals are nearly as curious as man, however, and also go caving. A few like it well enough to set up housekeeping. Some years after my last visit, I recommended Albright Cave, Washington, to a local Sunday school class. They had a wonderful time until they started out of the cave—and noticed that a bear was happily hibernating just inside the entrance. (They were a bit quieter leaving than entering, and now they know what a bear cave smells like! If you don't know, visit the zoo before spelunking in bear country!)

About a hundred miles further south, I once met a cougar strolling out of a lava tube cavern. It looked as curiously at us as we did at it, and since it walked away without even lashing its tail, we correctly concluded that there were no kittens inside and went ahead with the mapping of the cave. Without a similar exchange of compliments, however, I don't recommend this for caves that smell of cat. Recent newspaper accounts told of a well-shredded Mexican spelunker who staggered back to his home village after finding it necessary to strangle a jaguar he had accidentally cornered. Had it been a panicked cougar, I'd have bet on the cat.

Wolves may still den up in remote Canadian and Alaskan caves. Aside from Isle Royale in Lake Superior, none still exist in the conterminous United States, but I'd just as soon not tangle with even a coyote in a crawlway, or somebody's pet dog gone wild. Still smaller animals have notably sharp teeth at bay.

Cattle and other large herbivorous animals go deep into the twilight zone to get out of the hot sun; beware bulls and stampedes. Porcupines are no problem unless you have to crawl over one to get out of the cave. They huddle up and tremble when they think they can't run—or whatever a porky calls his waddle—any farther. And, if you glimpse something dark with a fluffy tail uplifted, and what might be a broad white stripe beyond, figure it's a skunk even if the cave is odorless.

VENOMOUS INSECTS

Scorpions and tarantulas live in entrance zones and deeper in tropical and subtropical caves; bees and wasps are not uncommon in cave entrances throughout much of North America. In my own home state, trogloxenic yellowjackets seem to specialize in stinging repeatedly and ferociously with one end while gnawing enthusiastically with the other. Yet none of these is as truly a cave hazard as the bite of the brown recluse spider, occasionally a cave dweller of the United States. Its venom is mild, but occasionally prostration accompanies severe pain. Black widow spiders are more aggressive, and, if their bite goes unnoticed, the symptoms may present a medical puzzle—sudden, seemingly spontaneous, prostrating abdominal pain or muscular cramps. This is usually relieved promptly by an injection of calcium gluconate, but the nearest hospital staff may not be accustomed to thinking in terms of unnoticed cave spider bite as a rare cause of excruciating pain in the abdomen, back, or extremities. Remind them, if necessary.

SNAKES

In the United States and Canada, venomous snakes and reptiles rarely venture past the entrance zone, but occasionally they do stray deeper, and may survive some weeks before starving or drowning. In caves supporting an extensive animal life, some boas of Latin America are dwellers. Venomous snakes occasionally become trogloxenes; the fer-de-lance has been reported in at least one Mexican cave. Apparently only pit vipers are cavers in North America, which simplifies the potential antivenom problem. These are said to emit a melonlike odor, which may help, since a startled rattler may not have time to complain about a caver's descending posterior.

Talus and cracks at the entrances to caves in "snake country" are a

Rattlesnakes are rarely found beyond the entrance of caves, but the author spotted this one trying to hibernate in Parks Ranch Cave, New Mexico, perhaps the largest gypsum cave in the United States. Note also the vadose speleogens.

different matter. Although tales of cavernous "snake dens" seem wildly exaggerated, particular care is worthwhile when entering and leaving. Texas cavers have whiled away many hours just inside small entrances, waiting for a rattlesnake to move on so they could leave.

BIRDS

Aside from the potential problems of bird-vectored diseases, mentioned later in this chapter, probably only one bird in North America is a hazard to the caver. A Texas acquaintance of mine has the unenviable record of twice having been regurgitated upon by cave-nesting buzzards—perhaps a more dire fate than being buried in a guano slide. Once he accidentally cornered an adult. On the other unforgettable occasion, a recently fed fledgling in a cavern-entrance nest defended itself needlessly but effectively.

And he's still an avid caver!

Other Medical Hazards

Especially for skilled spelunkers in good physical condition, most of the legendary risks of the mysterious netherworld are far less than the noncaver thinks. Yet certain spelean problems transcend first aid and require the caver to add something of the detective and the physician to his skills. Some are probably only theoretical hazards. Others have befallen cavers and will probably recur in the future, despite all precautions. Most, however, are preventable, or can be minimized with ordinary care and average knowledge.

SEWAGE DISEASES

Kentucky's Horse Cave is a revolting example of what a community should *not* do with sewage. Such fetid caverns, however, are so obvious a problem that, when faced at all, it is with due respect. Clear, odorless underground streams and seepage also carry bacteria and viruses for surprising distances under both karstic and pseudokarstic terrains. Early in this century such an infection partially crippled and almost killed the great French "father of speleology," Edouard-Alfred Martel. Today sewage diseases remain the greatest medical hazard to North American cavers.

In the recent past, the streams of mountain and other wilderness caves far from human habitation generally had low "natural" concentrations of bacteria and other germs. Except in floods (when bacteria counts usually skyrocket) it was safe to drink freely from such underground waters. With today's outdoors-oriented hordes practicing far too much "random elimination," locales where cave water is safe have shrunk almost beyond recognition. Even in remote regions, the chance exists that some animal died or was dumped in one of the next few sinkholes upstream. No longer do I drink from cave streams anywhere in North America unless I would drink from the same stream on the surface.

In temperate climates, the major hazards are the nonspecific but acute diarrheas that recently flushed a major expedition out of a Virginia cave. South of the Rio Grande, where these are generally termed "Montezuma's revenge," the problem is even more acute.

Infectious hepatitis is a rarer but definite hazard, and so perhaps is polio. In warmer climates, amebic and various bacterial dysenteries (including some rare subtypes of typhoid fever) must also be feared, as well as intestinal worms and other hopeful parasites. Cholera rarely

Rimstone pool in the New Mexico Room of Carlsbad
Cavern. Note beginning stalagmites at lower left.
Unfortunately, drinking from such pools is usually unsafe.

jumps the oceanic barriers, but if an outbreak occurs in karstic or pseudokarstic terrain, cavers had best suspend operations for many months. Typhoid and polio immunizations should be kept up to date. If the risk of infectious hepatitis appears significant, immune globulin may provide protection.

Animal feces are additional sources of disease. The greatest threat is tetanus, which is caused by puncture or other deep wounds infected by a germ that occurs naturally in the intestines of horses and other animals. Although often curable through today's medical technology, tetanus remains so truly a dread disease that every caver should maintain a high level of immunization against its effects.

The dumping of animal carcasses into caves and sinks causes contamination with different sorts of bacteria. Despite zealous efforts of public health authorities against such disposal, many a caver still finds himself in the local dogcatcher's refuse pit, or crawling through the entrails of a poached deer. Sometimes the bacterial concentration is so great that minor scratches threaten severe illness and permanent deformity, even in today's antibiotic era. It is the better part of valor to retreat immediately, wash extremely thoroughly with plenty of soap and hot water, and notify one's physician, so he can plan to begin appropriate antibiotics at the least hint of a serious infection. Surgical soap or Phisohex is a little better.

Especially in sheep-raising or caribou country, seemingly healthy cavers unexpectedly found to have one or more large, puzzlingly round spots on a routine chest X ray would do well to suggest to their physician that they may have echinococcus disease. This is a parasitic disease of sheep and a few other animals that is occasionally transmitted to man by dogs, wolves, and water. The spots are caused by cysts, which must often be treated by particularly complex surgical techniques.

Anthrax is also a potential threat. So far it appears to be only theoretical, which is fortunate—for this is an especially nasty disease even in cows.

Spelean trash and garbage disposal is easily considered. Anything carried into the cave can and should be carried back home—and a bit of somebody else's, too.

As for human waste, the same principle applies. Pack it out! Plan to enter caves with bowels and bladder empty and leave with them full. Before and during caving, avoid laxatives and diuretics like coffee, tea, alcohol, and excessive fluids of any kind. Leave nothing but footprints.

But be careful with what you pack out. A member of the NSS Southern California Grotto recently mistook a urine bottle for one that still was full of drinkable water. According to a local newsletter it took him a can of fruit and an entire chocolate bar to get rid of the taste.

CAVE-DUST PNEUMONITIS

Cave-dust pneumonitis is one of the three diseases inherent in some North American caves that are more specific medical hazards to spelunkers. "Dust pneumonia," as it is known to cavers, is a fairly common chest problem in dry cave areas. Caused by breathing considerable quantities of cave dust, it is quite like other types of dust inflammations of the lungs and bronchial tree. Rarely, however, is it mentioned in medical literature, and few cases come to the attention of physicians unless one is among spelunkers exposed to the dust.

The exact cause is not known. Cave dust is an extremely complicated mixture of many things that seem likely to bother sensitive lung and bronchial tissues: bat guano; excreta of rats and other animals; dried-up remains of nettles, poison oak, and innumerable other plants; various mineral salts; bacteria; molds; worm and centipede eggs; dirt; and quite a few other interesting items.

The caver usually first suspects that he has cave-dust pneumonitis a few minutes or hours after he has inhaled too much cave dust, when each breath begins to hurt more and more. Often he becomes slightly short of breath and develops an uncomfortable dry, rasping cough. The pneumonitis lasts only a few hours or a few days, and may thus be an undiscovered allergic or chemical response to some specific item in the dust. It is not due to powdered limestone itself, for the lungs consider this a remarkably inert material.

I personally have contracted cave-dust pneumonitis twice, in Bod-fish Cave, California, and in Garden Park Indian Cave, Colorado. Several cavers in other parts of the United States have sent me information on their attacks. None has reported any lasting result, but I can verify that it is thoroughly uncomfortable while it lasts. Archeologists have learned to wear dust masks in such caves, and cavers would do well to follow their example.

HISTOPLASMOSIS

Once described as "cave sickness," acute histoplasmosis is a more serious dust-borne disease of caves. It is a particular hazard in certain tropical and subtropical caves, but has afflicted cavers as far north as the middle Mississippi Valley. Many other individual cases and epidemics have resulted from breathing other dust contaminated with bat guano or bird or chicken excreta.

In contrast to cave-dust pneumonitis, which develops rapidly and painfully after exposure to dust, acute histoplasmosis has an incubation period of four days to more than two weeks. Its symptoms are rather like those of virus pneumonia, with chills and fever, headache, vomiting, irritability, prostration, a bit of a cough, and chest pain—which may be dull and constant or sharp, "cutting off each breath" (shortness of breath, however, is relatively mild and comparatively brief). Severe cases may lapse into coma. The physician finds little when he listens to the breath sounds, but X rays usually show a diagnostic pattern of innumerable fuzzy little patches. Skin and blood tests usually become positive after a few days of illness. Most victims begin to recover after a week or two, but the mortality rate is about 1 percent. Shortness of breath, weight loss, fatigue, and a minor fever may persist for several months. The X ray characteristically clears slowly, but instead may gradually change to a spectacular pattern of small, shotlike spots throughout the lungs. Complicated or very serious cases may need intensive medical or surgical treatment.

A somewhat similar dust-borne disease named "coccidioidomycosis" (and thus usually called "San Joaquin fever," or "valley fever") occurs in the warmer parts of central California, Arizona, and points south.

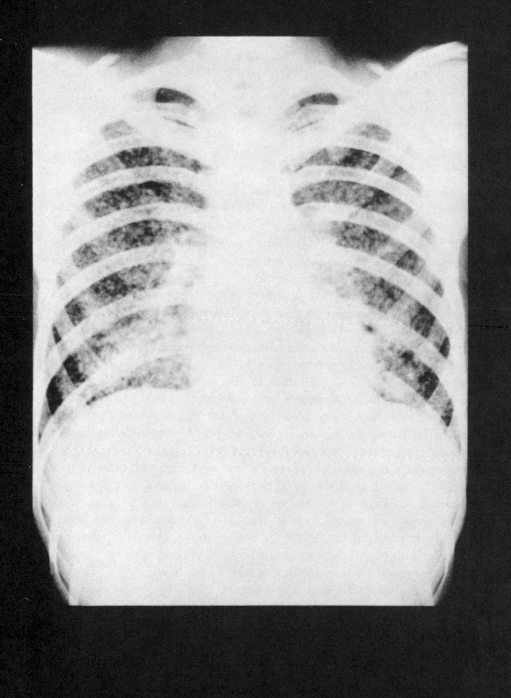

Limestone and lava caves bordering the San Joaquin and Sacramento valleys of California are on the edge of its range, and others in the American Southwest and Mexico are precisely within it. I have been told that at least one caver has developed valley fever from a spelean exposure, but no such case seems to be formally on record at this time.

RABIES

It has long been known that rabies (hydrophobia) can be transmitted by the bite of infected vampire bats, which are relatively common in Latin America and the South Caribbean area. The disease can also be transmitted by the bite of infected small mammals that inhabit caves throughout much of North America: skunks (four U.S. cases in the 1960s), foxes (thirteen such cases), coyotes, wildcats (one such case), and several species of the insectivorous bats that range north into Canada. Every decade, several persons are bitten by the latter—normally inoffensive, pleasant little winged animals. Despite intensive treatment, three persons in the United States died from this dread disease during the 1960s, and many more in Latin America. Fortunately, no caver has yet been among the fatalities.

The virus that causes rabies floats free in the atmosphere of some caves with huge bat colonies. Two persons working in these caves have died of rabies without being bitten; one entered only briefly. While both may have become infected through breaks in the skin rather than breathing virus, test animals exposed to the cave atmosphere consistently contract the disease. Active immunization against rabies is now available, and most spelunkers, speleologists, and biologists working with bats or otherwise exposed should undergo these injections. The vaccine is not as benign as those against tetanus, typhoid, and the like, but the risk is small in comparison to that of rabies.

Among unimmunized cavers, any bat bite in the Caribbean, Central America, Canada, and the United States (excluding rabies-free Hawaii and possibly Alaska) should be treated exactly like a bite by a dog in the same area. If someone can hang on to the bat until it can be

Appearance of a typical chest X ray of acute histoplasmosis, once called "cave sickness." The fuzzy white spots seen between the crisscross rib patterns are characteristic of the disease.

tested for rabies, and especially if it looks healthy, treatment can be deferred for a few days pending the results of laboratory tests of the bat. This does not apply if the bite is on the face or neck, where the incubation period is much shorter.

If the bat could not be caught, and especially if it was behaving erratically, preventive treatment should begin at once. Some healthy-looking bats are infected, and only now are physicians beginning to save the first few victims after signs of the dreadful disease develop.

Rabies is so fearsome a disease that various schemes periodically propose extermination of local cave bats. Accordingly, someone burns to death or otherwise slaughters thousands of them—a horrible tragedy to those of us who know bats as happy little animals that curl up in one's hand and purr. No one has ever explained what such slaughter is supposed to do to all the rabies-carrying bats that never go near a cave, or all the other animals that carry the illness. Much more appropriate is leaving bats alone, to be handled only by immunized biologists, and deferring the exploration and study of bat caves to immunized cavers. Even those who have no personal liking for bats should respect their inroads on the local bug population—especially in these anti-DDT days.

Rarer Diseases

Caves shelter animal vectors of other human diseases. Among those which might produce obscure symptoms are sylvatic plague, relapsing fever, Q fever, tularemia, tick paralysis, and Rocky Mountain spotted fever—a group of diseases transmitted by ticks, fleas, and mites that may be parasites of bats, birds, and cave rats. Weil's disease (spread through the urine of rats) is known to have befallen an American caver, and rat bite fever is a theoretical possibility, even though our cave rats are of the genus *Neotoma* (pack rat) rather than *Rattus*. American cave snails have not been recognized as disease vectors, but, especially in parrot-rich Mexican caves, cavern-dwelling birds might be a source of ornithosis and psittacosis. Mosquito-borne diseases may be encountered unexpectedly in midwinter, since some overwinter in caves. Reduviid bugs, which are a vector in a nasty form of trypano-somiasis called "Chagas' disease," have been found in Mexican and Central American caves, and probably occur more widely. "Swimmer's itch" (an annoying but trivial form of schistosomiasis) can be expected in cave waters where it occurs on the surface. It is prevented by prompt toweling. Bilharziasis (a much graver disease caused by *Schistosoma mansoni*) may be a serious hazard in caves of Puerto Rico, the Lesser Antilles, and perhaps elsewhere in tropical North America. Native lore in Puerto Rico insists that swimming is safe above waterfalls, but I wouldn't care to test it.

Contagious Diseases of Man

Especially in view of the high rate of contagious diseases among the dyssocial fraction of North American youth—some of whom are cavers—these must be considered here. The extreme intimacy of an exploring party under strenuous, fatiguing conditions results in vulnerability to tuberculosis and other respiratory contagions. "Atypical" tuberculosis germs have been found living in ordinary soil, and may occur in that of caves. Infectious mononucleosis can be transmitted by canteens. In caves, however, venereal disease transmission is solely by the usual route—for which most northern and high elevation caves are not suitable.

Emotogenics

Claustrophobia is a panicky fear of tight places. Its severe forms are rarely encountered by cavers, for most claustrophobics back off at the entrance. Although few of us will admit it, many cavers panic mildly each time we get thoroughly stuck or otherwise in trouble. Experienced cavers can almost always prevent the development of serious hysteria by conscious relaxing, which does much to correct being wedged or whatever the problem may be. Deliberately thinking a way out of the mess usually does the rest.

In the days before we knew about auto-belay Prusik knots, I once got my feet too high in an overhanging rappel and had to be "talked down," just as I had talked other panicky cavers out of nasty spots. The calm, studied technique was so obvious that I had to laugh, and my fear vanished.

Calmness, definiteness, matter-of-factness—such an approach almost always steadies the victim, even in the face of grave danger. Panic is contagious, demoralizing, and physically weakening. Hysterical or conflicting advice and irrational behavior must be controlled at once. Good leadership may demand strong distraction, including verbal countershock, or even force. Not long ago a falling rock glanced into a group of nationally known cavers, who should not have been beneath the cavern entrance. A girl fell, struck on the shoulder. I was a few yards down a slope, and another physician was almost as close at hand. Fellow cavers helped me rush to her, but her husband hysterically refused to allow anyone but himself to do anything. Had she been more than bruised, the incident could have been tragic.

Once the situation is controlled, the victim will usually return to normal, or at least to rational functioning, within minutes. Occasionally he or secondary victims of contagious panic must be evacuated from the cave as soon as possible. This especially befalls close friends of an injured caver, who may be in worse shock than the victim.

Illusions sometimes are a bit of a problem. On the surface, voice-like overtones of running water are easily recognized for what they are. Underground, whole parties of overtired cavers have further over-extended themselves seeking the fellow cavers they "knew" were close at hand. Usually someone retains his grasp on reality. While the others may not be immediately convinced by his exasperated insistence, they are likely to abandon the fruitless search much sooner if he keeps nagging them.

Hypothermia or exhaustion may even cause more dangerous delusions, subtly altering one's mental image of the return route. Where others recognize the correct turn, the affected caver may be so convinced they are wrong that he stumbles off alone. When this happens, the others must pacify the agitated, confused victim. It is best to stop for a few minutes, to check everyone for signs of hypothermia, and prepare some warm fluids. This usually settles the victim sufficiently that he is willing to try the group's decision rather than separate the party.

Mere startle reactions are sometimes unnerving. As I entered a narrow side corridor of a single-chamber cave on Utah's Promontory Point some years ago, a huge white-feathered ball of something—seemingly much too large for the little passage—exploded out of nowhere, right in my face. It left the cave faster than I did, but not much. Many minutes after I had identified it as a large, terrified owl, I was still shaking too much to finish mapping the cave. Practical jokers among Carlsbad guano miners are known to have gotten equal results. Fortunately, cavers don't play dangerous jokes underground. I suspect the rest of us would make stalagmites out of anybody who tried.

Accident Prevention

Accident prevention in caves goes a step beyond the obvious. Persuading the immature or overanxious to give up "showing off" or acting tough is an example.

Moreover, society recently has begun to realize how many cause their own accidents—accidents that are not truly accidental. Faced with seemingly overwhelming stresses, virtually any accident sometimes looms as a happy mirage, a seemingly priceless respite from the emotional pain of existence. Mood changes in one's caving companions sometimes foretell impending catastrophe. Increasing medical and psychiatric research on "the accident process" has revealed that an amazing number of accident victims were concealing despair, blind rages, or other self-destructive processes.

Professional help may be the only aid for these persons "looking for an accident." Confronting a disturbed person with one's concern is rarely helpful, and return assurances are false comfort. Yet at times the mere touch of a warm thought or any other easing of the victim's self-defined burden prevents untold tragedy. If nothing else can be achieved, efforts should concentrate on transferring the inevitable accident to the surface, for the sake of all cavers everywhere. In caving, the old ideal of "a healthy mind in a healthy body" may not be so old-fashioned.

Drugs and Medicine

STIMULANTS AND PSYCHOTROPIC DRUGS

Recent studies of psychotropic drugs have clarified their role in caving. Some were evident from the beginning of "the drug scene." Whether "on a bad trip" or otherwise, the dangers of hallucinating underground are so obvious that no reasonable person would go caving while under the influence of even the most popular hallucinogens. Yet no one has ever claimed that all cavers or all users of "street drugs" are reasonable, and sometimes the two groups overlap. I have had to help rescue one caver who tried to free-climb an obviously unsafe pitch after an undetermined amount of marijuana. Fortunately he lived, but the outcome was in doubt for a while. Problems as severe as temporary paranoia are known in people "coming down" from Methedrine or cocaine and few persons on "street drugs" know exactly what they are taking.

I have seen one "drug scene" individual surreptitiously gulp down what I assume was "speed"—Methedrine—when extremely tired and chilled. His movements soon became jerky and labored, and, although his recollections are to the contrary (like those of some afflicted with chronic cannabis poisoning), we had considerable difficulty getting him out of the cave.

That one observation doesn't mean much, especially since I needed help in getting out myself. But medical evidence is accumulating an impressive consensus that can be summarized in one word: *Beware!*

"ENERGY DRUGS"

Controlled testing of athletes given test doses of Methedrine, Dexedrine, and other amphetamines has shown that these drugs increase rather than decrease fatigue, as is widely believed. Sympathomimetic drugs (those with an action resembling adrenaline) do produce a short burst of energy, but at the expense of longer-range performance. At least four athletes have died from misuse of these drugs, and a few developed dangerous delusions during supposedly safe tests.

Thyroid extract and anabolic steroids like testosterone are useless in caving; their long-term actions are potentially even more serious than those of the sympathomimetics. They do not build stamina as legend hopefully suggests, nor even short energy bursts. Even in the case of mixed-up cavers who seek to hide dismaying hangovers until they are far underground, caffeine appears to be the only safe stimulant. And that only in the form of tea or instant coffee or medically prescribed dosage.

VITAMINS

On cave trips of less than forty-eight hours, vitamin pills are of no value to any caver who normally eats anything resembling a balanced diet. A chronically malnourished caver facing a long, desperate rescue struggle of many hours or days may be helped by supplemental basic vitamins. Even then, however, several days will probably be necessary before he reaches normal. Use of vitamins E or B_{12} is a valueless fad.

But belief can lead to improved performance. Another group of athletes were told that they were to test a new "energy pill"—actually nothing but a tiny, insignificant amount of sugar. Performance improved 63 to 72 percent after they took the pill.

ANALGESICS, NARCOTICS, AND DEPRESSANT DRUGS

Virtually the only drug cavers should give each other for pain is aspirin. Don't snicker, for recent, sophisticated pharmacological testing has shown that aspirin is at least as good as other analgesic drugs except narcotics, with one possible exception—Talwin, which is a little stronger and can be given by injection when the stomach rejects all pills. This new agent, however, may soon come under narcotic regulation, as did Demerol a generation ago, so it appears best to standardize with aspirin. Fifteen grains (two or three tablets, depending on the size) every three hours or so until the stomach rebels or the ears begin to ring, with as much water as the victim can manage. A little bread, milk, or some other bland food may help an irritated stomach, but should not be given if vomiting is likely.

The Bureau of Narcotics and Dangerous Drugs is undoubtedly aware that many establishment-type mountaineers have carried small supplies of narcotics for years. This is technically illegal, but there has been no abuse, nor any known objection or interference. Establishment-type cavers look and smell almost as bad as drug-culture cavers, however, so I cannot recommend that they do the same. Just as in all other matters of underground first aid, the byword is clear-cut: it's better not to become a statistic!

Barbiturates (phenobarbital, Nembutal, Seconal, etc.) are likely to cause a disabling drug hangover—about as bad as an alcohol hangover— if taken in a misguided attempt to insure a restful sleep the night before caving. These and all other depressant drugs (including alcohol and marijuana and its derivatives) cause a subtle reduction in efficiency that can be critical underground. Although psychotic reactions to these drugs are uncommon, they do occur, and accidental withdrawal must also be considered. A silly statement may be the only warning of an incredibly irrational action: walking blindly into an obvious pit, or starting up a wall of tottery breakdown. Victims of such reactions can be evacuated only under immediate supervision so that they have no opportunity to harm themselves or others. If the reaction does not subside promptly, a rescue team must be called.

Physical Fitness

Despite the rest of this chapter, the medical aspects of caving have a bright side. Proof is fast accumulating that caving is a healthy sport. Keeping in shape for caving and other strenuous activities is increasingly recognized as an excellent way to help ward off heart attacks. Recent studies on exercise by the American Heart Association are likely to benefit out of-shape cavers far younger than my rear-spreading generation. The AHA concedes that regular, vigorous exercise "tailored to the capacity and interest of the individual" runs a calculated risk in some population groups—including those with middle-age flab. Its Committee on Exercise, however, has found that medical benefits of getting fit and staying fit far outweigh this calculated risk. Furthermore, the committee has gone a step farther and developed a conditioning system by which the risk can be minimized.

As we all know, fitness rapidly deteriorates when we refrain from exercise. It now appears that the type of conditioning is relatively unimportant as long as it leads to sustained increases in function of the heart and lungs, and of body metabolism in general. A little exercise is much better than none, but it must be on a regular, graded basis if the caver is to stay fit or get back into shape—three times a week to get fit and twice a week (not weekends alone) to stay fit. For vertical cavers, Bill Cuddington especially recommends jogging and the Coopers' *The New Aerobics*. For those over forty, or with known medical problems, or currently or recently on drugs (including alcohol), exercise should be under careful medical supervision.

Well known also is the need to start at a low level and build gradually. Warmup and tapering-off periods should last three to five minutes or more. Not so obvious is the committee's finding that fifteen to twenty minutes regular *appropriate* exercise is often all that is necessary. Data cited by the American Medical Association's Committee on Exercise and Physical Fitness indicate that such exercise should progress to 70 to 80 percent of the individual's maximum effort. The AHA has a guideline of not pushing the heart rate past 75 percent of its maximum in the early stages of conditioning or reconditioning.

Healthy young cavers can normally define their own maximum without much difficulty. For the middle-aged flab and higher-risk groups (like survivors of one or more "coronaries"), the AHA has investigated and endorsed special testing programs, which determine just how much exercise is likely to be safe for each individual. For the fittest over forty, the maximum heart rate appears to be somewhere well under 140, ten or more points higher in younger healthy cavers. If your heart seems to jump rapidly three times consecutively before returning to the normal pulse, it means danger; taper off immediately, rest, and consult your physician.

This is greater progress than the average caver may realize. One of the leading explorers of the Summit Caves of Mount Rainier, at an altitude of more than fourteen thousand feet, has heart disease of a type with which he was born and characteristically causes no problems until later in life. Until recently, he would have been forbidden such caving. Before I switched from surgery to medical administration and rehabilitation, I sometimes took patients caving less than a month after major chest surgery. Yet until now I would not have dared prescribe caving for someone rebuilding his life after a heart attack. Whole new concepts of rehabilitation speleotherapy are unfolding. If you want to be a caver, make regular physical activity your way of life.

10

Cave Search and Rescue

November 4, 1967, was a busy night for Terry Tarkington. As nearly as I can reconstruct it, his part in the action began about 7:30 P.M. The place: Huntsville, Alabama, later to become official headquarters of the National Speleological Society. The event: a telephone call from Troy Watts near Anvil Cave. For several hours a search had been under way for five boys in that maze of all mazes—almost thirteen miles of cave beneath some eighteen acres of sink-pitted pastureland!

Knowing the cave well, Terry responded immediately. Nearing the cave area at 8:20, he encountered the tail end of a stupendous traffic jam. Later he recorded selecting "a virgin, four-wheel drive route around [it]." In plain English, he drove clear around the mess. The fun was about to begin.

8:25. Terry locates and announces to the National Guard, the Flint Rescue Squad, and assorted bystanders that they can now relax. The Cave Rescue Unit (himself) has arrived and will have the boys out within a few minutes. Blank looks, no relaxing—but no interference: the power of positive thinking.

8:26. Terry looks around for competent help and recognizes Jimmy Moore, a local high school footballer who has been in the cave a few times. Terry hands him a hardhat and light and the duly constituted search party is off.

8:30. At entrance number one of Anvil Cave, Terry and Jimmy encounter twenty or thirty weeping parents and other kin of the lost boys. Patiently they explain that there are no pits or other particular hazards in the cave and they will be back in minutes. Credibility is greater than at rescue headquarters.

8:35. Terry and Jimmy enter the cave. Immediately they encounter "a mile of string [in] the first couple of hundred feet inside the cave.

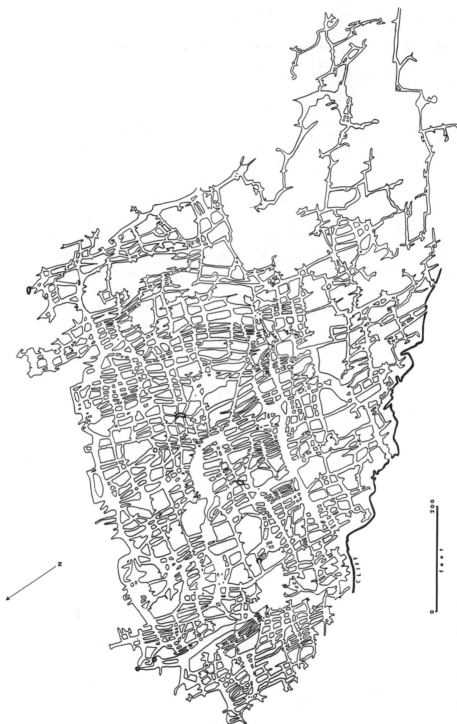

Fig. 20. Map of Anvil Cave.

We had to fight our way through it, crossed and crisscrossed in every direction as though we were in a gigantic spider web."

8:40. Terry and Jimmy encounter the human spiders—several would-be rescuers with a single flashlight and a large ball of twine.

8:45. The "spiders" are convinced to return to the entrance. Terry and Jimmy break out of the net of string and set off at a fast walk, shouting at each junction as they "sweep" the cave systematically.

8:52. The lost are found.

9:00. The lost are restored to a hysterically happy crowd. Reporters interview Terry and tremendously overdramatize the event.

9:05. Somebody mentions that there are still two search parties in the cave.

9:10. Terry and Jimmy go hunting again.

9:20. One missing search party is found. Its members deny vigorously that they are lost.

9:25. The search party is escorted to the surface through a convenient sinkhole entrance.

9:30. Terry and Jimmy return to headquarters. The other group is still missing, but someone suggests that everybody wait awhile—which is okay with Terry and Jimmy, who have covered a lot of cave in a hurry.

11:00. The missing are still missing, so Terry and Jimmy go after them.

11:20. A long sweep of the cave locates them far to the southeast.

11:30. Everyone out of the cave.

11:45. Coffee with the Moores; sure tastes good.

Midnight. Homeward bound from a fun evening.

Not every cave search has a Terry Tarkington who knows the cave like his own home. Six months earlier three boys had vanished from the face of the earth near a similar Missouri cave they had been exploring. Despite weeklong search operations of incredible extent, they remain missing to this day.

Unfortunately, the Missouri episode illustrates some of the problems of cave search better than the farcical Anvil Cave "rescue." If the cave is not intimately known by someone at hand, or if the missing persons are not found promptly, or if they are skilled cavers who are likely to have penetrated obscure orifices, the search may be extremely difficult. The Anvil Cave episode illustrates the use of small, fast-moving parties to sweep likely areas. If they find no trace of missing explorers, full-scale operations must be initiated immediately. Regional and national cave rescue organizations should be notified, and extensive efforts begun to obtain information on any other cave where the missing spelunkers might have gone.

Organization of such a search varies little from that of a rescue (described later in this chapter), for some of the missing spelunkers may be injured. The cave is again swept, spreading inward from the most likely to the least likely areas. Novices are likely to be in or near a large chamber entered through a small hole they cannot find when they want to go home. Or where the wrong turn in a network would lead them. Experienced cavers are likely to be in a remote, even virgin cave.

Search parties must anticipate a lack of answer to their shouts. Many a boy lost in a cave is petrified at the thought of his parents' wrath. Others have been silenced more tragically—by rockfall, drowning, or falling.

While the second sweep is getting under way, consideration of alternates begins. In a 1955 California fiasco, the missing caver was located in New York—he hadn't bothered to tell anyone when he changed his plans. Probably he still cringes at recollection of the black headlines: UC STUDENT LOST IN CAVE, and so on. So do those who went looking for him.

A major search of a cave includes preparation or revision of an accurate map showing progress in each area. Special investigations can be color coded: rock piles, underwater search, digging, and the like. A series of maps showing areas studied in ever-increasing detail may help direct efforts. When all else fails, a more recent map should pinpoint features that might have changed since the cavers' passage: rockfall, rising water, mud slides, and the like. Not all missing cavers have the good sense to stay where they are, and may wander back into an area already swept. Periodic resweeps may help.

Searching for a Cave to Be Searched

The calls that especially try cavers' souls are the vague ones. Someone can relate only that some boys went to look at a cave somewhere west of Schmoesville and haven't come back. . . . No, the hysterical parents never knew there were any caves west of Schmoesville. Neither did local cavers. Nor the sheriff or anybody else you can find.

Or a couple of happy-go-lucky cavers went somewhere and never came home. (If you're lucky, the sheriff will soon find their car in an area with only a dozen known caves to be swept.) Because the chance of finding missing cavers alive decreases rapidly in much of North America, great gobs of time, effort, energy, and money often must be channeled into innumerable lines of investigation, even if it's probably a false alarm.

The national, regional, and grotto files of the National Speleological Society contain a wealth of information that can be tapped for such emergencies in the United States. They also contain valuable information on caves elsewhere on this continent, although not as systematically. The National Cave Rescue Commission will do everything possible in such a situation, as will the office staff of the National Speleological Society (see list of equipment and information sources). Unfortunately, such information tends to be slow to reach the national files.

Except in Cuba, no organization similar to the National Speleological Society exists elsewhere in North America, but several regional and local Canadian groups are among NSS affiliates. The Association for Mexican Cave Studies has extensive files, and maintains some contact with Mexican cavers. It too usually can be contacted through the NSS or the National Cave Rescue Commission. The Sociedad Venezolana de Espeleología maintains liaison with Spanish-speaking speleology in general. Its address is Apartado 6621, Caracas—which doesn't help much in an emergency. The staff of the Escuela de Geología of the Universidad Central de Venezuela in Caracas (especially Professor Franco Urbani) may be a better source. Russell Gurnee (231 Irving Avenue, Closter, N.J. 07624) is a bit of a one-man Guatemalan speleological survey. In Belize, cave files are maintained by the Departments of Archaeology; in Costa Rica, by the Departamento Parques Nacional and the Grupo de Espeleología del Club de Montañeros de Costa Rica. Geological surveys like that of Jamaica may be helpful.

Information can often be obtained from landowners, government agencies, and other sources spelunkers normally tap to learn about new caves. Occasionally local law enforcement offices like those of the Royal Canadian Mounted Police know of caves that have eluded local speleologists. And district rangers and administrators in a variety of bureaus and departments, and mountain rescue units, are also well informed.

U.S. and other Geological Survey maps, county and other "legal" maps, and those of units of government occasionally provide clues. The files of local libraries, colleges, and outdoors clubs should not be overlooked, nor should querying the manpower sources mentioned later in this chapter. At last resort, radio and television stations will broadcast pleas for information. Every responsive report of a cave must be investigated under these circumstances, for only the unexpected is routine. Many a cave has turned up where experts scoffed.

A guide is infinitely superior to the most detailed directions to a cave. A California ranger once gave me excellent directions to an important but remote cave shown on some road maps. From about two miles away, he even pointed out which tree the entrance hides behind: "You can't miss it!" You guessed it. Eight hours later we stumbled back down the canyonside, defeated. As usual, too many trees, all looking the same. If you can find someone who has seen the entrance of the missing cave, take him along!

For those accustomed to reading the language of geomorphology, that language may prove the only guide—an imperfect guide, to be sure, but much better than none.

Hope must never be given up until—well, until everyone must give up all hope. The brotherhood of cavers demands no less.

The Rescue Operation

As January ended in 1925 near Mammoth Cave, the search wasn't the problem. A solo caver named Floyd Collins had not come home one Friday night. Alarmed friends found him within minutes next morning, "trapped three hundred feet underground."

Coast-to-coast suspense gripped the United States as rescuers vainly tried to lever a "six- or seven-ton boulder" off his foot. (Actually Collins was less than seventy feet underground and "the boulder" weighed only 27 pounds, but they had to work through a tiny hole alongside his cramped body.)

The heart of the American people went out to Floyd Collins, and perhaps killed him. Thousands of would-be helpers impeded rescue operations. Blundering crowd control by the authorities soon excluded virtually everyone with first-hand cave experience. Even earlier, some would-be rescuers had panicked, covertly abandoning the vital supplies that were staving off the hypothermia that doomed him. Hysteria and drunkenness ran rampant, and inexperience caused the rockfall that sealed his fate. When an ill-considered rescue shaft was begun, it was in the wrong place. And Floyd Collins, only slightly bruised, almost certainly was already dead.

Yet this oft-retold 1925 tragedy was not wholly in vain, for it taught America how cave rescues should *not* be conducted.

Sometimes we remember.

Fortunately, solo cavers like Floyd Collins are largely specters of the dim past. Today the crucial question is whether others in the victim's party can rescue him with an adequate margin of safety for all concerned.

As indicated in the last chapter, an injured caver who cannot be evacuated readily should be kept safe, warm, and comfortable until a large, experienced, and well-equipped party can do the job safely.

Ideally, that is. Cave rescues are rarely in ideal surroundings. In much of the colder parts of North America, rapid, highly sophisticated evacuation must be accomplished by those in or close to the cave if the victim is to survive. In warmer areas, protection from hypothermia and shock are more effective. There, evacuation can often be postponed.

The victim's fellow cavers must plan rapidly and realistically. They must balance the shifting risks constantly, asking themselves if they can further delay evacuation—and, if so, how long. Sometimes the victim's reaction and condition force reconsideration almost from minute to minute. More often, the nature and severity of the injury,

Fig. 21. Sketch of Mammoth Cave and the three flat-topped ridges beneath which it extends. (1) Cascade Hall, reached from Flint Ridge by CRF explorers after passing beneath the intervening valley and all but the farther slopes of Mammoth Cave Ridge itself; (2) Historic Entrance of Mammoth Cave; (3) Floyd Collins Crystal Cave Entrance; (4) Austin Entrance; (5) Entrance to Salts Cave; (6) Colossal Cave section; (7) Unknown Cave section; (8) George Morrison's New Entrance; (9) New Discovery entrance (artificial); (10) Great Onyx Cave; (11) Proctor Cave entrance; (12) Turley Cave (Morrison Cave) section. Shaded ridge-top outlines based on U.S. Geological Survey topographic maps, cave passages simplified from maps by National Park Service, Max Kaemper, U.S. Geological Survey, Cave

the difficulties of the situation, and the availability of rescuers pre-determine the decision.

Either way, a small party can often accomplish much more than they think in the first shaky moments. If someone has a single small pulley along with ropes and carabiners, forethought and ingenuity can work wonders. Even if outside help is clearly needed, the victim's group can do much: heating water, moving rocks, damming streams, emptying pools, and the like. Such activity helps stave off hypothermia among those who have donated much of their clothing to their needy friend. It also benefits the morale of all concerned.

Decisions to call for help must be made as soon as the need is clear. Yet this should not be done lightly. Cavers' needs are much like those of mountain rescue organizations: one telephone team, one rescue leader, forty-eight stretcher bearers.

Actual cave rescues are likely to require much more complex organization. A typically difficult rescue may well need:

1. One telephone team (communications director near the site of the rescue, coordinator of a regional cave rescue organization's team, telephone committee of that organization, national cave rescue coordinator, telephone committee of local civil defense units, and various other local organizations and agencies)
2. One rescue director at the cave entrance
3. One assistant rescue director, usually with the victim (may or may not be the helper on vertical pitches)
4. One press secretary or information officer
5. Two to six belayers
6. Two to forty haulers and stretcher bearers, often largely recruited from local civil defense and mountain rescue organizations

Obviously, no one wants to set such an operation in motion needlessly. Yet it should be done without hesitation as soon as it is evident that those immediately at hand may not be able to evacuate the victim safely.

MESSENGERS

When outside help is likely to be needed, two members of the party should be dispatched to seek *and control* it as soon as they can be spared, and one of the pair should be a particularly experienced caver. Until specifically relieved by someone he recognizes as competent, he will serve at the cave entrance as temporary rescue director (use of formal titles is invaluable in resisting well-meant takeover attempts by incompetents).

The victim's life is likely to depend on the safety of these two messengers. No matter how experienced, they must be nagged to slow down, consciously, to a safe pace of all due deliberate speed, for they run a high risk of additional disaster.

Each should carry a copy of specific information on waterproof paper:

1. Precise location of the accident, with detailed directions to the cave and victim
2. Name, address, and condition of the victim
3. Type and severity of the injury
4. Names and addresses of those remaining with the victim, their condition, and their tentative plans
5. Distances (a) from the cave to a road and/or helicopter landing site; (b) from the victim to the entrance
6. Obvious problems and obstacles, and special needs (hypothermia, crawlways, pits, cliffs [underground and surface], special folding stretchers, ladders, ropes, physician, cave divers, and so on)

To these must be added:

7. The address and phone number of the location from which the call is being made

The messengers must not merely transmit this information into the nearest telephone. They must make sure that it reaches competent assistance, with as little alarm and as little publicity as possible. The name and address of the victim and the name and location of the cave, for example, should be especially guarded. The Floyd Collins debacle was not the only cave rescue hampered by milling mobs of curious bystanders, incompetent volunteers, and overzealous journalists.

The temporary rescue director prepares the site for the drastic activities that will soon be converging. Where possible, roadblocks should be established a mile or more from the cave. Specific areas should be marked or roped off for various operations, including a convenient location where journalists can observe and inquire without interfering. A register must be created to log in and out everyone entering the cave. Ideally, equipment should be similarly logged.

EXPANSION OF COMMUNICATIONS

Even at this stage, planning the expansion of rescue communications pays off. Often this is best accomplished with the assistance of a police radio car, citizens' band or "ham" radios. Occasionally a telephone line may be needed to connect the accident site, the cave entrance, and the police car or nearest regular telephone. This may be beyond the abilities of the local telephone company, but it never hurts to ask at the proper moment.

If a language barrier is likely, the more bilingual messenger should begin as communications director, then shift to temporary rescue director if appropriate.

LOGISTICS

As assistance begins to arrive, more and more planning is necessary: medical aid, food and drink for the rescuers, electricity, road and helicopter access, traffic control, and equipment inventory and storage, and appeals for special tools like hydraulic jacks. Also necessary are purchase and safe storage of dynamite if the entrance must be enlarged (liaison with local police may be very helpful), liaison with the Weather Service for short- and long-range forecasts; appointment of the press secretary to provide accurate, nonsensational news—not opinions or speculations—to the working press, and to gain their cooperation in return; and much else, including detection of those who are so jittery that they must be kept busy. Much of this can be shifted to the first experienced, competent caver who arrives. Normally he takes over as rescue director for the next few hours. No arguments can be permitted about who should be rescue director, and the temporary director must serve until someone arrives to whom he gladly relinquishes the operation.

THE NATIONAL CAVE RESCUE COMMISSION

The National Cave Rescue Commission can provide advice plus reasonably quick assistance throughout much of North America. It functions best when provided enough time to place key personnel on standby status. Therefore it should be kept informed even when it appears initially that local resources are sufficient. Its 24-hour telephone number is 1-800-851-3051.

OTHER SOURCES OF HELP

If the NCRC number is out of order, try the NSS office, or the U.S. Air Force Rescue Coordination Center, Scott Air Force Base, Illinois. Or call the information desk or emergency telephone at Mammoth Cave, Carlsbad Caverns, or Wind Cave national parks. If you have an amateur radio operator available, try the mountain and cave rescue frequency: 155.16 MHz. Keep trying. Meanwhile, start mobilizing local resources. The local telephone operator and postmaster, however, should *not* be asked unless all else fails (the word spreads far too far, far too fast).

Where they exist, mountain rescue and civil defense organizations should be the first noncaver contact. Technically, these usually cannot respond until requested by "duly constituted civil authority." In practice, they know how to get the job done.

Next on the list of choices is the closest major law enforcement agency. Many are skilled in surface search and rescue activities, and most of the others do surprisingly well. When necessary, almost all will grit their teeth and venture into what they correctly consider dangerous black chasms.

In remote areas, governmental agencies are often well prepared: the Coast Guard, the National Park Service, district rangers of the U.S. Forest Service, the Bureau of Indian Affairs, the Bureau of Land Management. In Canada, the Parks Department and other national and provincial administrators are of help, as is the army in Mexico and some other countries.

Chance-met hunters, fishermen, loggers, and other outdoorsmen should not be overlooked, nor should construction workers, truck drivers, the staffs of commercial caves and of nearby colleges (especially the geology and biology departments). Most will be happy to help and happier to leave underground leadership and technical skills to competent spelunkers.

As I've indicated above, the communications director is likely to be a novice, so every caver has a clear-cut duty to his fellow cavers. Before each enters any cave, every member in the party should be agreed on the proper sources to call for help.

But people being human, it won't always happen. And all the suggestions above are subject to failure. If all else fails and the situation is getting out of hand, one last-ditch procedure may be necessary: telephoning the highest government executive whose staff can be reached—after telling the press that you are about to do so, and why. This is a desperation move. The caller must be prepared to face bitter resentment, for he must charge total incompetence or vicious lack of cooperation. The complications of this will be virtually unmanageable, and the rescue operation will probably grind to a temporary halt. But this may be necessary to save the lives of more than the victim.

ORGANIZATION OF TEAMS

As competent rescuers, willing manpower, and vital equipment pour onto the scene, specific teams begin to attack specific needs of the moment. Often, only the leaders of such teams need be experienced

Carroll Slemaker in a Stokes stretcher after suffering a compound leg fracture in Church Cave, California.

cavers. Antihypothermia teams are equipped with plastic tarps, warm clothes, blankets, food, heat sources, and other emergency gear. Communications teams with radios and telephone sets, equipment teams to scout and rig the evacuation route, and to transport splints and stretchers are also needed.

HORIZONTAL EVACUATION

If the victim can be placed in a Stokes or other rigid stretcher and carried to safety, the operation may be virtual routine to search and rescue organizations—merely a tedious, backbreaking, totally exhausting race against time. Many caves, however, limit the use of such rigid stretchers. A wide variety of collapsible stretchers is likely to arrive, and, whenever a more collapsible model arrives, it should be expedited inward.

If someone can recall the simple technique under the emotional strain of the moment, a surprisingly useful stretcher can be woven from a 120- or 150-foot rope. It begins with two small loops near the middle of the rope, about eighteen inches apart, just big enough for comfortable handles. Similar loops are tied every eighteen inches for approximately seven feet in each direction (a total of three or four loops). These form carrying straps.

The stretcher-to-be is then laid on the ground and the open end is closed with a square knot. The free rope ends are laced through the loops, then diagonally back and forth. As they pass another length near the center of the fast-forming stretcher, frequent twists or knots add stability. If necessary, sling ropes may be added to make a more uniform bed.

"Basket stretchers" can also be woven by combining standing and sling ropes. While these can be used in hoisting a victim, they are trickier to construct. Those wishing to perfect them may refer to standard mountain rescue manuals.

CARRYING THE VICTIM

Where no stretcher is available, the easiest way to carry an injured caver is "piggyback." Unfortunately, the size and shape of the cave and the nature of the injury often fail to cooperate. If the patient has an injury much above the knee, and especially if he is in shock, some other technique must be devised. If the bottom doesn't rip out, perhaps he can be carried in a large pack, his legs dangling through holes cut in the bottom. Or a figure-eight rope coil, if he can tolerate the upright position.

The over-shoulder "fireman's carry" illustrated in many first-aid manuals is easy for the rescuer, but so hard on the victim that its usefulness is limited. The four-hand carry, on two rescuers' linked wrists, can be used for short distances. If the victim must be carried lying down, two or three pairs of rescuers are needed. Each pair links wrists beneath the victim, carrying him in lockstep. Much more tiring is alternating rescuers on each side of the victim, each cupping his body in their arms or on their shoulders.

All carries are so tiring that any kind of stretcher is a great benefit. If long poles are available, they can be rigged to permit dragging the victim as well.

CRAWLWAYS AND SQUEEZEWAYS

Crawlways are not impassable barriers to rescue. Some must be enlarged by sledgehammers or blowtorches (which convert limestone to slaked lime) but expert use of manpower often renders this needless. A team of six to eight wrist-carriers can duck into a stoopway and go on until exhausted (which will be soon). Another team takes over, then another, until the first team is again ready. Short obstacles may require rotating carriers around the victim, with the rescuers on the obstructed side dropping back as those on the opposite side advance and cross over beyond the obstacle. Sometimes only a single carrier need drop out. Surprisingly often, the victim can assist enormously by a small motion at the proper moment.

Still tighter crawlways may necessitate laying the victim atop a rescuer, who then crawls as best he can with assistance from others. Sometimes it is necessary to drag the undercarrier as far as his clothing and skin permit (this works best in streamways, where buoyancy is a favorable factor). If possible, a crude travois of sticks, lath, heavy cloth, rope, or almost anything else should be substituted for the hapless underman. Where the crawlway drops or rises abruptly, a caver's body or a quickly heaped pile of rocks or packs simplifies the effort.

Narrow squeezeways may require carrying the victim on the rescuers' locked arms, shoulders, or even heads. The latter cannot be long tolerated unless a rigid stretcher is available. If some of the rescuers can work at a higher level, they may be able to support much of the load with ropes while one or two proceed beneath the victim.

WATER PASSAGES

While water is obviously a grave problem in some cave rescues, it may be useful. If the victim can be kept dry, almost anything floatable will reduce the rescuers' burden: planks, waterwings made of plastic parkas, small inner tubes, other cavers. Occasionally damming small streams creates a helpful temporary pond.

PIT RESCUE—SIMPLE HAULING

Pit rescue can be remarkably quick and easy, or one of the most difficult of all cave problems.

Perhaps the simplest technique involves a four-wheel drive vehicle with an adequate winch. The rescuers rig a Swiss seat and ascender box or other secure harness and haul him out. Especially if the ascent is against a wall, cable spin can be moderated or controlled by a second caver ascending with the victim. A taut belay several yards from the winch may also help. The jeep's brakes must be set, the gears in reverse, and its wheels blocked. Cavers may snicker at the idea of the victim, the helper, and the jeep all going together, but it could happen.

Lacking a winch, a victim can be hauled up by manpower. If both ropes are nonlaid, belaying may be feasible. Where the rope comes over the lip and runs horizontally, abrasion must be prevented. A carabiner pulley rigged across the pit or two or three feet above its lip or some other additional hoisting rig may be a great help. The haul rope must be protected continuously with Prusik or ascender. If properly rigged, the rope runs smoothly while the haulers are completing each tug, then is automatically held by the knot or ascender. One rescuer should be stationed at each safety device, however.

In this method, special danger lurks at the lip. When the victim is ready to be helped out of the pit, the rescuers are likely to forget their own exposure. Each person approaching the edge must be anchored or belayed. Waist belts are a valuable reminder.

The value of nonstretch ropes like Bluewater II is evident during pit rescues. Ideally, ropes of different colors minimize confusion, but they soon are usually all the same color: mud. Numbering is a poor substitute. People forget which is which. Names are better: haul rope, standing rope, victim belay, helper belay, counterweight rope, and so on. All should standardize the names, for communications are likely to be poor.

And it must always be remembered that rescue operations cause far greater stresses on ropes than normal caving—as on cavers.

PROTECTING THE VICTIM

As a result of bitter experience, mountain rescue operations recommend that victims be lashed in stretchers by not less than six individually secure slings. In the absence of a stretcher, a seat sling plus a chest sling may be used.

If part of a stretcher ascent must be out of the horizontal, the victim must also be anchored to its head. If he faces even momentary suspension by his armpits, these must be padded extremely well to prevent arm paralysis. Thick padding everywhere plus a face visor of some sort (the belayer's helmet?) will do much to protect him from the falling rocks that are often loosened by the inherent clumsiness of rescue operations.

POSITIONING

The difficulties of easing a helpless caver over the lip can be greatly reduced by rigging the haul rope over a pulley well above the lip. Thus, with a little slack, he can be swung in for a soft landing.

In some locations this is not possible. If the victim can tolerate a momentary upright or sitting position, sling ropes attached to the stretcher or pack at about hip level may be used to alter his center of gravity, thus permitting him to be tilted over the edge. If not, another caver alongside or below him may have to boost him.

A somewhat similar problem may arise in narrow, irregular chimneys. If the victim can tolerate short periods in the head-up position, an extra rope attached to the head of the stretcher permits much-needed maneuvering. Another from the foot or middle of the stretcher to a caver below may also help control spin and gain maneuverability.

THE HELPER

Even with the victim lashed securely, adjusting the center of gravity is usually difficult if the stretcher must be kept horizontal. If it strikes a tiny knob, an unaccompanied stretcher may career wildly. To steer it away from such projections, to control spin and tilting, to reassure the victim, and to treat midair problems like vomiting, hoists should be accompanied by a helper.

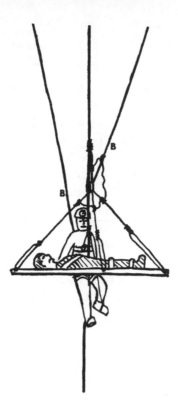

Fig. 22. Diagram of safety slings when raising victim and helper on a single rope. All but one of the stretcher's side ropes and many unrelated details like the face shield have been omitted for clarity. Note that both helper and stretcher are belayed. Helper is attached to haul rope by three-knot system. Stretcher is suspended from haul rope by short sling of 7/16-inch or several thinner sling ropes. Each stretcher rope is "safetied" at both ends, as is short sling connecting haul rope and stretcher carabiner. Slack shown here (for clarity) should not be permitted in actual use. Belay ropes are indicated by letters B.

Normally the helper is whichever rescuer is most fit. He may ascend as a unit with the victim, or separately by ladder or rope. If manpower is extra plentiful, it may be possible to haul him up on a separate rope. Ascending as a unit with the victim has many advantages, but it does roughly double the load and often is not feasible.

Especially if wall walking is possible, the victim may be able to ride on the back or shoulders of the helper, well anchored to the rope and preferably facing the same direction. Regardless of the method chosen, victim and helper are best belayed separately.

If the helper is ascending by ladder, he usually functions most efficiently two or three feet above the victim in horizontal stretcher hoists. If the victim is rising in the vertical position, the top of the stretcher should be somewhere around the helper's thigh. Presumably much the same is true if the helper is ascending a separate standing rope. When ascending as a unit on a single hoist rope, the helper's slings are placed above the attachment of a horizontal stretcher so that it virtually rides in his lap. If the victim is head up, the top of the stretcher normally rides about his chest level. In all cases, his position is likely to be subject to change without notice.

PULLEY SYSTEMS

If haulers are scarce but a long rope is available, the mechanical advantage of a pulley system is invaluable. Simplest is a rope anchored somewhere topside, running down to and around a pulley securely attached to the victim or stretcher, then back up to the hauling crew. Such a pulley should be rescue or haulage type rather than the simpler carabiner pulley, which may produce more friction than the haulers can handle. The precautions mentioned earlier in this chapter (i.e., being anchored) are necessary at the lip. As in all pulley systems, nonstretch ropes are preferred, and the rope must be considerably longer than twice the depth of the pit.

If large, high-quality pulleys are at hand, still more mechanical advantage can be gained by running the anchored rope down the pit to a double sheave securely attached to the stretcher. Thence the rope runs back up to a second pulley, back down to the second sheave compartment, and thence up to the hauling crew. For this the rope must be more than four times the depth of the pit. The presence of four pulley lines adds considerably to the risk of rope snarl.

A solo rescuer or a team can make effective use of a simple traveling pulley system created with a shorter rope and two rescue-quality pulleys. The haul rope passes through the first pulley, then doubles

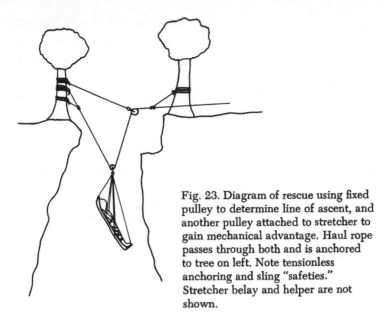

Fig. 23. Diagram of rescue using fixed pulley to determine line of ascent, and another pulley attached to stretcher to gain mechanical advantage. Haul rope passes through both and is anchored to tree on left. Note tensionless anchoring and sling "safeties." Stretcher belay and helper are not shown.

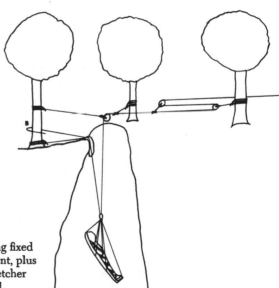

Fig. 24. Diagram of rescue using fixed pulley to determine line of ascent, plus traveling pulley (at right). Stretcher belay rope passes over rope pad.

back through the second pulley before continuing to the haulers. The second pulley bends the rope in a Z, but causes little damage even to stiff-fibered ropes because of the low coefficient of friction. (Or so I am told by the experts. Not having been a math major, their calculus intimidates me.)

This second pulley is also hooked to the haul rope, above the lip, by a short sling and ascender knot or device. (As in all hauling, an anchored ascender or knot is placed even closer to the lip, as a fail-safe precaution.) As the rescuers pull, the pulley attached to the haul line by the short sling moves along the rope toward the other, producing mechanical advantage. If the two pulleys touch, the Z comes out of the rope and the advantage is lost. This is an annoyance rather than a problem, since it is easily replaced. After each haul, the rope is supported by the anchor below the Z while the traveling pulley is being repositioned.

Carabiners are poor substitutes for pulleys in these systems, but can be used in emergencies.

COUNTERWEIGHTS

When a speedy rescue is especially vital, a counterweight is often the fastest and safest method. Even if no special urgency exists, this technique may be helpful if manpower is limited and only a pulley or two is available. By using a counterweight, a lone rescuer can raise an injured comrade in several ways.

The simplest counterweight technique employs a load slightly heavier than the victim. This is anchored securely to the end of a rope, which passes upward from the victim and through a pulley anchored above the lip. Separately anchored safety slings are emplaced on both sides of the pulley, and adjusted with each haul if necessary. The counterweight may be a bagful of rocks, a plump fellow caver, maybe even a duffel bag full of water (if it can be kept from emptying itself). Especially if rocks or other loose objects are used, a second pulley should move their fall lines well away from the victim.

In adjusting the weight of the counterbalance, the ideal is reached when the rescuer must use a slight pull to raise the victim. When the

Fig. 25. Rescuer climbing in place with small counterweight (pack). Note anchor on haul rope below pulley and face protection. Rescuer can be belayed or anchored.

two weights are about equal, he should be able to move the system in either direction by a pull of about twenty pounds. With a rescue-quality pulley, a twenty-pound advantage by the counterweight still permits friction to bring the system to a halt, yet minimizes the effort of the rescuer above. If the only available counterweight is much heavier than the victim, its descent can be controlled with a rappelling device; if much lighter, the rescuer must expect to haul until exhausted, rest, and try again. If another caver serves as the counterweight, soaring to the bottom as the victim rises, he does most or all the hauling for the first half of the job. This is a risky technique, for he can be belayed only on fairly short pitches, and *only with kernmantel rope.*

A more flexible counterbalance technique combines the weight of a caver with a comparatively small counterweight. With this technique, a rescuer can "climb in place," securely anchored, while the victim rises toward him. Or he can descend to any chosen point to assist the victim and control the stretcher as needed. While ordinary standing rope techniques are used, smooth, gentle motions are desirable in order to minimize stresses on the system. The rope walker can be belayed with a kernmantel rope for a moderate distance, but dangerous rope snarls are increasingly likely as additional ropes go farther and farther into the depths.

All counterweight systems exert unusual stress on ropes and anchor points. Often such stresses are several times the weight of the victim. The inherent insecurity of pitons and other artificial anchor points and slings may require substitution of some other technique.

CROSSING PITS

At least three methods permit transporting a helpless victim across a wide pit. The tiniest of ledges may permit a simple traverse, requiring insertion of many pitons or bolts and a fixed rope or two, plus elaborate belays for all involved. A helper is usually needed at each end of the stretcher, and sometimes another beneath it, scrambling along a nearly nonexistent ledge in true hodag style.* Usually the stretcher must be belayed from each end.

Aerial ropeways are often discussed but rarely feasible. These are modifications of mountaineers' spectacular Tyrolean traverses along a taut nonstretch rope. Unfortunately it is almost impossible to get enough sag out of ropes for this to be efficient. Nevertheless, this may be the best way to transport a stretcher through a seemingly upside-down cave like West Virginia's Schoolhouse Cave, which is floored by

* A hodag is a mythical cave animal with short legs on one side for walking along slanting walls. Some have suction cups instead of toes.

pit after pit. Pulleys suspend the stretcher to minimize friction. Two haul lines as long as the taut rope are needed. Only occasionally can a belay be used effectively, and usually only to slow the stretcher on the downhill section of its course so that the ropes do not snarl.

Treating a wide pit as a descent plus an ascent on the far side is often the easy way to pass it. The team must judge rather precisely the distance to be descended against one wall before swinging to the other for the ascent. Often it is best to descend all the way to the bottom.

RESCUING WEDGED EXPLORERS

The only way to learn if a crawlway "goes" is to push it until the smallest, most agile caver gets stuck. (Some hold that caving is considered a team activity to ensure that others are at hand to uncork such situations.)

Theoretically, two basic concepts are applicable when someone becomes thoroughly stuck: pulling the hapless cork regardless of painful minor injuries or protecting him from flooding, hypothermia, ants, and other annoyances while waiting for him to slim down a few pounds. Delightful tales recount alleged examples of the latter, but most or all are mere yarns. Hypothermia, however, is the real foe, and not even the U.S. Bureau of Mines seems to know how soon rescuers' efforts may become a lost cause.

Some have tried applying great gobs of grease to wedged explorers. Sometimes it helps. A favorite cavers' tale of the northeastern United States recounts such an episode in the 1950s. Stuck fast for several hours, a well-known caver from a distinguished university finally popped out, "clad in a thick coat of grease and nothing much else," according to a gleeful chronicler. Those present seem pleased to recall that the victim nobly upheld the honor of all concerned. According to legend, he bowed formally to the grinning crowd, apologized for his attire, and expressed due thanks.

Falling down a narrowing fissure or pit may be even more critical than being trapped by a rock no one can reach. The force of the fall wedges the caver deep and tight, and his chest is likely to be so severely compressed that heart and lung action are endangered. Especially if he is wedged sideways, both victim and rescuers must work rapidly and effectively for him to survive. Should he wedge head down, he must be pulled out immediately. Even the upright position may quickly prove a fatal trap.

Such locations are likely to be comparatively poorly ventilated, so a long air hose and oxygen tanks may be needed urgently. Welder's oxygen is just as good as medicinal. And just as explosive. Any open flame (including that of carbide lamps) near flowing oxygen is likely to cause an explosion, with the tank shooting about like a crazed rocket. Compressed air is safer and often easier to locate. It is much less effective, however, and probably should be used only until oxygen arrives. Small, lightweight oxygen supplies are available at many drugstores and surgical supply houses, but weight is a minor consideration under these circumstances.

If the caver is on belay, one's first thought may be to try pulling him out by the rope. If he wedged at any speed, however, this is likely to bunch his clothing and skin and make matters worse. If he has an arm free, lowering something he can grasp is preferable. Normally he should merely grasp while others pull. Attempting to pull one's self upward in such circumstances tenses and bunches the muscles and tightens the wedge. Ideally, he should grasp a short sling tied to a pulley dangling on a rope loop.

If the rescuers can reach the trapped caver, cutting his clothing away may be a key to escape, but also to hypothermia through heat conduction.

Brute strength is rarely the answer, for such victims are wedged incomparably harder than the stuck crawlway-crawler. Pulling on the arms dislocates shoulders, even tears limbs off the body under such strains. Perhaps better is rocking the victim back and forth while pulling gently, or moving him sideways toward a more spacious section or a foothold. A tiny toehold may make the difference between escape and failure. If the victim's blindly groping feet are unsuccessful, it may be possible to swing a rope or loop where he can hook it with one foot and push down against it with the other. A cable ladder may be even more effective. Once he reaches a point where he can chimney perceptibly with his toes, then his knees, the worst is probably over.

Fatal Accidents

The sad need to remove a human body from a cave has led several notable cavers to quit the sport. Yet this tragic responsibility cannot be left to noncavers. While every situation is different, certain guidelines are clear. The remainder of the dead caver's party should be evacuated before their shock and grief cause further tragedy. If possible, the body should be recovered by cavers who did not know the victim personally. The leader of the recovery party must anticipate and preclude additional shock, especially among younger, more excitable members of his grim team. All should be steeled for a revolting, grisly scene. The corpse should be covered and removed as rapidly as possible without endangering anyone. As long as an injured caver lives, his fellows gladly face tremendous risks for his sake. At death, this changes totally.

The body should be wrapped as thoroughly as possible to protect others from emotional shock. If nothing else, the face, head, and especially bloody and gruesome areas should receive all possible attention from the most mature and stable member of the team. Transportation is simplified by tying the legs together, and the arms to the sides. The hands should go inside the trousers, or in pockets. Often the recovery is simplified by dragging the body feet first.

Notifying the survivors may be the worst part of the tragedy. Sometimes this is best done by a noncaver—in any event, by a compassionate person and preferably by one dear to the survivor. The most grisly underground task pales in comparison.

For the problem of the wedged caver who cannot be rescued, the body that cannot be safely recovered, there is no right answer. Yet some wrong answers are less wrong than others. All should be proud to say: "I did the best that could be done."

FLOYD COLLINS TODAY

The type of problem that faced Floyd Collins may be best approached indirectly. Had he been so trapped today, teams of skilled cavers would pour into Mammoth Cave and others nearby, seeking a passage that would permit them to come up behind him and lift the fatal rock off his leg. (Mapping parties of the Cave Research Foundation have passed almost exactly beneath Floyd's trap, and no large cave is ever completely explored.) Others would begin to tear away rocks in nearby orifices that might "go." The bulk of the rescuers probably would begin to move the seventy feet of cliffside rubble that trapped Floyd, and it's anybody's guess which rescuers would be successful.

Alternately, an antihypothermia team might find it necessary to administer an opiate and drag him out by brute force (Floyd's brother Homer came near doing this until his brother's unnarcotized agony became too great). Then and now, amputation was no solution. No surgeon could reach the proper level. Similarly, glib talk of deliberately fracturing the collarbones of an injured caver to fit him through a small hole is largely bravado. It might help sometimes, somewhere, but I doubt anyone doing it.

Cave Hunting

Closely allied to searching for missing cavers is ordinary cave hunting. When cavers travel, they just naturally look around for good caving areas, yet in vast areas caves and congenial caving companions may be difficult to locate.

Obviously, it's nice to find someone who knows nearby caves and talks your language, and the seeming barrier of conventional languages often intrudes less than many would believe. In broad areas of the United States, Canada, and Latin America, cave hunting from scratch is needless.

Here the experienced explorer has a special advantage. Much more likely than not, he is a member of the National Speleological Society. Often he need only review its list of chapters—"grottos"—and regional organizations to find new caving friends. The novice can begin similarly by writing that society at Cave Avenue, Huntsville, Ala. 35810. Should that address change, the current one can always be obtained from the American Association for the Advancement of Science, of which the National Speleological Society is an affiliate.

In areas without grottos or regional associations, the society's an-

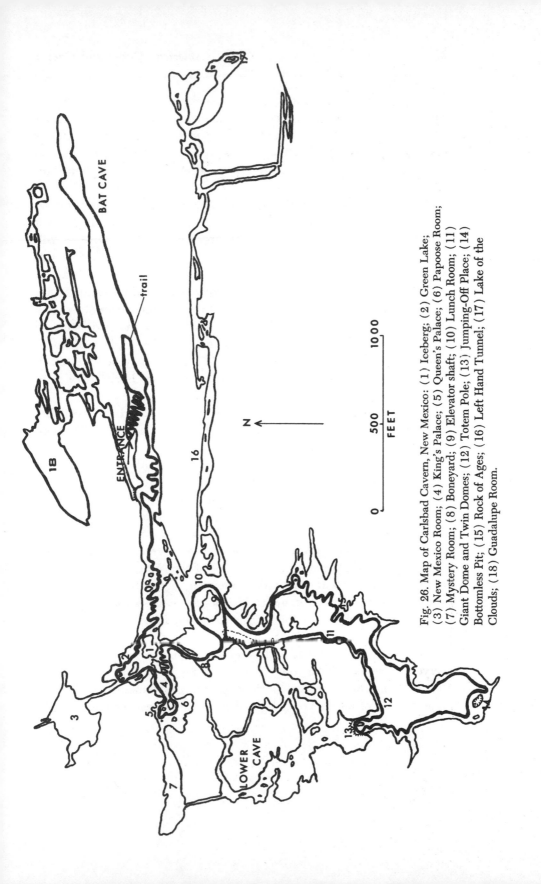

Fig. 26. Map of Carlsbad Cavern, New Mexico: (1) Iceberg; (2) Green Lake; (3) New Mexico Room; (4) King's Palace; (5) Queen's Palace; (6) Papoose Room; (7) Mystery Room; (8) Boneyard; (9) Elevator shaft; (10) Lunch Room; (11) Giant Dome and Twin Domes; (12) Totem Pole; (13) Jumping-Off Place; (14) Bottomless Pit; (15) Rock of Ages; (16) Left Hand Tunnel; (17) Lake of the Clouds; (18) Guadalupe Room.

nual list of many thousand members usually suffices. Membership in the National Speleological Society has so many benefits that the serious North American caver can hardly afford to be without it, no matter what his country: publications, regional and national meetings and field trips, training courses, library, visual aids, cave and photo files and address lists, and much more.

But not even the National Speleological Society has members everywhere, and many of the caves of even the United States are not yet recorded in any file or publication. In broad areas, the newcomer will have to seek caves and cavers in the ways mentioned earlier in this chapter.

And under a significant handicap. Obviously the time factor is not as important as in a rescue operation. But the assistance volunteered so eagerly in an emergency is less enthusiastic and more difficult to locate. None of the sources listed above should be neglected, but a different approach is often worthwhile. Much time may be invested in libraries, thumbing through books and files that are likely to refer to local caves.

Nor—let's face it—are the various files on caves nearly so available if no emergency exists. Consider the cave files of the National Speleological Society, which contain more accumulated data on caves of the United States—and some areas ɔeyond—than all other sources combined. The sheer bulk of these files overwhelms the volunteers who man them. They are basically for the use of members of that society, and, except in emergencies, priority is given those members who contribute consistently to them. Certain staff members of research organizations and accredited institutions also receive preference. So do survey teams, regional organizations, and grottos of the National Speleological Society. Then comes the individual member, but only occasionally is manpower available to respond to his specific requests for information. Almost never if it is "all the information you have on caves of the southern part of Blank State." Some day, perhaps. Not soon.

When a caver learns that he has fellows in an area he plans to visit, everyone does best if he contacts them as far in advance as possible. Most National Speleological Society grottos and Canadian and other organizations of cavers are delighted to welcome responsible spelunking visitors—especially those with pictures and stories of their own favorite caves back home. Hospitality has some limit (we lost track somewhere around forty-six visitors the summer of the Seattle World's Fair). But, given enough time, virtually every caver is happy to make enjoyable suggestions to and special arrangements for his visiting brother.

Some cavers are hesitant to seek others, fearing that they may not be welcomed, or congenial, or have any way of returning the hospitality, or have sufficient experience. My advice is to go ahead and contact the National Speleological Society and its local units anyway. Nearly all are warm, friendly, understanding, and sympathetic, for most of us had similar doubts not long ago. In most parts of the United States and Canada and much of the rest of North America, almost anyone interested in caves and caving is welcome without obligation—except to respect caves and fellow cavers as we do.

If distance or national boundaries are a problem, the National Speleological Society can often link the visiting caver or novice with some other caving organization, or at least an individual of like interests. If that, too, fails, local offices of the U.S. Forest Service, National Park Service, Bureau of Land Management, or Parks Department may be able to offer good advice on what caves to visit, and provide the addresses of local cave enthusiasts. Commercial cave staffs are usually less helpful. They have enough troubles already, as sketched in my *Depths of the Earth*. Don't pester them.

Mexico is a little different. For most gringo cavers, getting along while south of the Rio Grande is a bit of an art. Liaison with the Association for Mexican Cave Studies is normally the first step. For caving in the Caribbean, the Department of Geology of the University of the West Indies (Kingston, Jamaica) and the Sociedad Venezolana de Espeleología (see list of equipment and information sources) may be able to help.

And then there are the neighbors. People who live close to caves. If you've figured out where caves ought to be, go ask—pleasantly, of course, and looking, speaking, and smelling like someone they would be pleased to assist. Always offer to be of service in return; carry their groceries from their truck to the house or help with any other current chore that might lessen their ability to think about caves. If you've read the face of the land and the heart of its people, you'll not only find the caves but make warm friends for life.

Protecting Your Welcome and That of Your Fellow Cavers

The beginner and the touring caver alike have a special responsibility to their fellow cavers and the caves: prevention of vandalism and injury chains.

Beware the good friend or chance acquaintance who has other, less responsible friends. They in turn may have bottle-smashing slobs for friends. Slobs who think it great sport to paint their names on cave walls in huge red letters. Fun to hear rows of delicate stalactites tinkling on the floor. Fun to attack colonies of helpless bats with clubs or torches. Fun to hide someone's rope while he's down a pit, or cut a nick in it just for fun, or even to pull it up and sell it.

Or maybe just careless slobs whose nauseatingly ripe bodies you may have to help haul out of a pit you happily explored the previous weekend.

Both types of tragic chain have occurred, even through the miscalculation of cavers especially dedicated to conservation and safety. They will occur again and again unless every caver proudly accepts his responsibility of educating his fellow man. Wherever you go, whatever you write, speak out persuasively on conservation and safety, safety and conservation. Persuasively, not demandingly. The old catch phrase about honey catching more flies than vinegar was shrewd indeed.

The touring and local caver alike share further responsibilities: recognition and furtherance of the brotherhood of cavers. Free to do his own thing where it affects no one else, each caver's every action and communication must bring credit upon every caver. In other words, courtesy and respect to every fellow human who identifies him as a caver. Adjusting to the slower life style of many who live in cave areas. Preservation of the scenic, scientific, wilderness, recreational, and other values of each cave wherever it may be. Finding ways to leave each cave and its environment the better for his coming. Discovery of underground water for Texas ranchers. Risky rescues of prize calves and mangy, snapping mongrels (dog-sized burlap bags can be tied to a Jumar). Photos and maps for cave owners. Reports to administrators of federal, state, provincial, or private lands. Courteously requesting permission to enter each cave, and courteously accepting refusals.

Even if he privately thinks the need outrageous, the concerned caver obtains any necessary permits and licenses, and complies with even the most stupid restriction invented by some distant noncaving theoretician—like permits for collecting bugs, licenses for radios, licenses *and* permits for the use of dynamite or stream-tracing dyes (some people panic when their drinking water unexpectedly turns green or red). If no discount is offered at commercial caves, he quietly pays the full admission price and snickers only inwardly when the guides are more entertaining than they know.

If he chances upon the bones of ancient animals or Indian relics, he shields them until science can evaluate their meaning through patient, systematic study. He treasures even the most repulsive-looking animal life as his own, knowing that it is even more fragile. He cherishes the brittle glory of even the least of our decorated caves as an integral part of nature, to be valiantly defended until the natural destruction of all things on earth shall come in its appointed time. Truly he takes nothing but pictures, leaves nothing but footprints, kills nothing but time.

Especially when muddy, tousled, and stinking with sweat and guano, he does his utmost to persuade the public that cavers aren't some weird breed of nut (even if he personally believes that we are!)—always remembering that the public long ago learned much of the silent language of appearance and action that all too often belies the noblest words.

Not being God, none of us achieves such continuous perfection. Few even aspire to it. Yet the caver who works consistently and effectively for the brotherhood of cavers is welcome virtually anywhere, anytime. As he develops ever-increasing proficiency in caving skills, his is truly the freedom of the world of caves.

May it be yours forever.

Glossary

AA. A rough type of lava in which lava tubes do not form.

ABLATION. The combination of processes by which glaciers shrink and glacier caves enlarge.

ABRASION. Mechanical surface wear.

ACETYLENE. The inflammable gas burned by carbide lamps as a result of the chemical reaction of water with calcium carbide.

AGGRADATION. Accumulation of stream-transported fills.

ANASTOMOSES. Intricate systems of tiny, irregular, twisting, flat-bottomed tubes that follow a single plane.

ANGEL'S HAIR. A somewhat fanciful name applied to clumps of unusually delicate gypsum needles that angle outward and curve slightly.

ANGEL'S WING. A gracefully folded type of drapery. Also applied to dripstone hanging from a vertical palette.

ANTHODITE. A confused term, originally applied to an unusual group of helictitic speleothems, but subsequently applied to other complex speleothems.

ANTICLINE. An upward arching of rock strata.

ARAGONITE. A mineral found in some caves, chemically composed of calcium carbonate, usually in the form of needlelike crystals but also found as stalactites, helictites, and other speleothems.

ASCENDER BOX. A metal housing for two pulleys, worn high on the chest in some ascent systems.

ASCENDER KNOT. A local name for the *helical knot.*

ASCENT KNOT. See HELICAL KNOT and PRUSIK KNOT.

AUTO-BELAY. (1) Use of a Prusik knot or related technique during rappelling in such a way that it will tighten and hold the rappeller in place if released; (2) a special rockclimbers' device providing a dynamic delay in case of a fall.

BACON-RIND or BACON-RIND DRAPERY. A thin drapery with bands of color mimicking a huge strip of bacon.

BAD AIR. A vague term usually applied when a caver suspects or knows that the air of part or all of a cave may be dangerous.

BALL, LAVA. See LAVA BALL.

BAROMETRIC. Pertaining to the constantly changing pressure of the atmosphere.

BASALT. A common type of lava. Aa and pahoehoe are forms of basalt.

BEDDING PLANE. The surface between two layers of sedimentary rock.

BELAY. Knowledgeable use of a safety rope.

'BINER. Affectionate abbreviation of *carabiner*.

BIVOUAC. A planned or unplanned underground camp with only gear readily carried in the party's packs.

BLIND VALLEY. A karstic valley drained underground and thus enclosed on all sides.

BLOCK-CREEP CAVERN. A long, narrow cave close to and parallel to the face of a cliff, resulting from the cracking away and "creeping" of a block of rock.

BOWLINE. A slipproof climber's knot.

BOXWORK. A complex of intricately intersecting thin blades of calcite or other mineral, projecting from the bedrock of a cave.

BRAIDED. (1) A term applied to lava tube systems that branch and rejoin one or more times; (2) a type of rope weave, quite unlike the laid or twisted type exemplified by Goldline. The surface of all kernmantel ropes is braided; their cores are either braided, parallel or corded.

BRAKE BAR. A short metal bar with a hole at one end whereby it is threaded onto a carabiner or rack, and a groove at the other that fits snugly against the other arm of the rappel device.

BRAKE BAR RACK. See RAPPEL RACK.

BREAKDOWN. Any material that has fallen from the ceiling or wall of a cave, but usually applied to considerable accumulations. Also used as an adjective, describing cavern chambers or other features formed or heavily modified by the process of breakdown.

CALCAREOUS. Pertaining to lime and related materials—limestone, calcite, etc.

CALCITE. The commonest of cave minerals, forming most speleothems observed by the average caver. Chemically it is composed of calcium carbonate.

CANOPY. A ledge or remnant of false floor festooned with stalactites.

CANYON. A deep, narrow stream channel in the floor of a limestone cave, often twisting. When passage-sized, the term "canyon passage" is sometimes used.

CARABINER. Metal "snap rings" used in many ways in caves. Some have a locking screw gate, others a simple gate that opens under direct pressure. The latter are increasingly considered unsafe for cave use.

CARBIDE. Calcium carbide, a solid chemical used as fuel for miners' lamps. See also ACETYLENE.

CARBONIC ACID. The weak acid resulting from interaction of carbon dioxide and water.

CARDIAC ARREST. Cessation of the heartbeat.

CAVE. A natural cavity below the surface of the earth, large enough to enter, with some portion in essentially total darkness. The term is often used more loosely.

CAVE BUBBLE, CORAL, COTTON, HAIR, ORANGE, RAFT, etc. Descriptive terms for speleothems more or less resembling the specified object.

CAVE ICE. Ice naturally formed in a cave. Sometimes used incorrectly for a delicate type of rimstone or shelfstone.

CAVE MILE. Technically, 5,280 feet of underground passage. Humorously, any distance underground over 100 feet or so.

CAVER. One who explores caves.

CAVERN. Same as cave. Sometimes a mild connotation of grandeur.

CAVE SYSTEM. An interrelated, basically continuous complex of caves, often separated by impenetrable segments.

CAVING. The exploration and/or study of caves.

CEILING CHANNEL. A distinct channel dissolved or gouged upward into the ceiling.

CENOTE. A steep walled sink of the Yucatan Peninsula area.

CHERT. A very hard, flintlike rock that occurs in beds or nodules in some limestones.

CHIMNEY. (1) Any opening more than about one foot in diameter leading upward in a cave—more specifically, one that is rounded and lacks the characteristics of a domepit; (2) a narrow, tubular volcanic pit; (3) to ascend or descend any narrow orifice by employing both walls as climbing surfaces or by pressure against both walls.

CLASTIC. Pertaining to rocks and particulate matter transported by gravity or stream action.

COLLAPSE CHAMBER. A cavern chamber formed or heavily modified by breakdown.

COLUMN. A compound speleothem produced by the fusion of a stalactite and stalagmite. Cf. PILLAR.

COMMERCIAL CAVE. A cave with an admission charge. Paths and other improvements are usually present.

CONCENTRIC. A large natural bull's-eye of congealed ripples on the floor of certain lava tube caves. Also used for smaller "bull's-eyes" occasionally seen on the walls of some limestone caves, usually where large coralloids have been subjected to resolution.

CONDUIT. A roughly circular or oval subterranean passage that serves or has served to conduct large volumes of water or lava.

CONFLUENT. Joining, in the manner of a tributary stream.

CONGLOMERATE. A rock composed of fragments of rocks, naturally cemented together.

CONSERVATION. Protection of ecological, wilderness, scenic, recreational, scientific, and other resources and values of caves.

CORALLOID. A small, nodular speleothem, usually of calcite or lava, often occurring in intricate complexes. Also termed "cave coral."

CORE. (1) The inner portion of a kernmantel rope; (2) the vital inner organs of the body.

CORRIDOR. A comparatively long, level, moderately straight position of a cave.

CRAMPONS. Mountaineers' "climbing irons" affixed to the boot for better purchase in steep snow and some ice climbing.

CRAWL or CRAWLWAY. A cavern passage too low for stooping.

CREVASSE. A glacier fissure.

CROSSOVER. A major passage that follows a course partially at an angle to another below it.

CUPOLA. An arched feature of the ceilings of some lava tube caverns.

CURTAIN. (1) A broad, wavy drapery; (2) a long row of intermingled stalactites.

DEAD CAVE. A cave in which the speleothems are no longer moist and enlarging.

DEHYDRATION. Loss of water.

DENDRITIC. Uniting in a pattern suggestive of the veins of a leaf.

DISTRIBUTORY COMPLEX. The pattern of some lava tube systems, characterized by downhill branching and rebranching.

DOLOMITE. (1) A sedimentary rock somewhat like limestone but less soluble because of a considerable proportion of magnesium carbonate; (2) a mineral composed of magnesium carbonate.

DOMEPIT. A roughly circular natural shaft in limestone or other soluble rock with sheer, slightly ribbed walls. Usually several feet or yards in diameter.

DRAPERY. A thin, pendant speleothem, often convoluted.

DRIPSTONE. Any stalactite, stalagmite, or other speleothem formed through the action of dripping water or lava. See also FLOWSTONE.

DRY SUITS. Flexible divers' waterproof garb.

DUCK UNDER. A point where an explorer must "duck under" a low spot to get from one place to another. Some are water-filled; see SIPHON.

DYNAMIC BELAY. A type of belay in which a fall is controlled by halting the run of the rope gradually, through a distance of several feet.

EFFLUENT. Branching away from another passage. Compare with CONFLUENT.

ELECTROLYTE. A chemical that ionizes in solution, or a solution or paste containing such a chemical, thus transmitting electricity.

ENSOLITE. Trade name for a plastic foam used especially for ground insulation.

EPSOMITE. Natural "epsom salts." Chemically, hydrated magnesium sulfate.

EVACUATION. Removal of an ill or injured person from a cave.

EXPANSION BOLT. A rock climbers' tool. After a hole has been drilled into the rock, the bolt is inserted. The bolt expands "into" the rock. Other hardware is attached to it.

FALSE FLOOR. A thin layer of flowstone, lava, or other material that conceals a space below.

FAULT. A plane or zone on which a block of the earth's crust has been displaced.

FIRN CAVE. A cave in snow that has not compacted sufficiently to develop the characteristics of a glacier.

FISSURE. A narrow crack in rock. Often used loosely for a narrow passage.

FLAKE. A slab of ice that peels away from the wall or ceiling of a glacier cave.

FLOW GROOVES, LEDGES, LINES, MARKS, etc. Descriptive terms for features of lava tube caverns with that general appearance.

FLOWSTONE. A surface coating of mineral, usually calcite or ice, deposited by a descending film of water. See also DRIPSTONE.

FORMATION. (1) A geological reference to a specific unit of bedrock; (2) a confusing popular term for speleothem, and more.

FUMAROLE. An outlet for volcanic gases, occasionally cavernous.

GEOTHERMAL. Pertaining to the internal heat of the earth.

GEOTHERMAL CAVE. A cave produced by geothermal melting of snow or ice.

GIBBS ASCENDER. A popular cam-type ascent device.

GLACIER CAVE. A cave in or beneath a glacier.

GLACIÈRE. Same as ice cave, but also including cold-trapping sites of other kinds, such as mines.

GLACIOSPELEOLOGY. The study of glacier caves and related phenomena.

GOUFFRE. A French term, sometimes applied to certain North American pit caves.

GOUR. Another French term, increasingly applied to rimstone deposits.

GROTTO. (1) A small side chamber of a cave; (2) a cavernous opening that does not extend into total darkness; (3) a chapter of the National Speleological Society.

GROUND WATER. Water in cavities or porous rock or soil. Some authorities exclude water in the subsurface zone of aeration.

GUANO. Speleologically, the excreta of bats.

GURNEE CAN. A tapered cylindrical metal can in which gear can be conveniently dragged through crawlways.

GYPSUM. A sedimentary rock and mineral composed of calcium sulfate, softer and more soluble than limestone.

GYPSUM BARREL, COTTON, CRUST, FLOWER, GRASS, HAIR, PLATE, ROPE, SAND, etc. Descriptive terms for various gypsum speleothems.

GYPSUM CAVE. A cave formed in gypsum, usually by much the same processes that produce caves in limestone. The term is occasionally misapplied to caves containing gypsum deposits.

HACKLING. A group of jagged little projections of bedrock, resulting from irregular solution.

HALAZONE. A nasty-tasting but effective chemical for killing bacteria in drinking water.

HALITE. The mineral form of common salt.

HANGER. A piece of hardware connecting expansion bolts to the next item, usually a carabiner.

HARDHAT. A caver's helmet.

HEAT TAB. A flammable chemical tablet.

HELICAL KNOT. A special ascent knot, really a simple pipe hitch.

HELICTITE. A vermiform speleothem.

HOMEOTHERMIC ZONE. The inner zone of caves, characterized by little if any perceptible fluctuation in air, water, and bedrock temperature.

HONEYCOMBING. A pattern of phreatic solution characterized by an irregularly rounded, intricate three-dimensional pattern.

HUM. A local Puerto Rican term for *mogote*.

HUNTITE. A complex mineral of magnesium and calcium carbonates.

HYDROLOGY. Speleologically, the study of underground water and its actions.

HYDROMAGNESITE. A mineral consisting of hydrated magnesium carbonate.

HYDROSTATIC PRESSURE. The pressure applied by a body of water above a given point.

HYPOTHERMIA. Significant lowering of one's core temperature.

ICE AX. A long, hatchetlike mountaineers' tool used for controlling slides on steep snow and for chopping steps in snow and ice.

ICE CAVE. A cave in which ice forms, and persists through much or all of the summer and autumn.

IGNEOUS ROCK. Rock of volcanic origin.

IMMERSION SUIT. Heavy rubberized one-piece garb originally designed to keep the wearer dry when floating for long periods.

INCHWAY. A crawlway so tight that explorers must force their way along, seemingly inch by inch.

JOINT. A crack in bedrock, caused by movement of the earth's crust or other natural processes.

JOINT POCKET. A prominent rounded alcove or dome, oriented along a joint.

JUMAR ASCENDER. A precision rope-gripping device for standing rope ascents.

KARST. Topography characterized by sinking streams, sinkholes, caves, and similar features indicative of underground drainage developed through the solution of bedrock.

KARSTLAND. An area of prominent karstic features.

KARST PROCESS. The process of base leveling through surface and sub-surface solution and corrasion.

KERNMANTEL. A type of rope consisting of a central core surrounded by a woven sheath.

LAID ROPE. A common type of rope consisting of twisted strands.

LAPIES. A term applied to shallow solutional grooves and channels on rock surfaces above and below ground.

LATERAL RIDGES. Residual features of a lava tongue whose center has slumped and/or gradually lowered.

LAVA BALL, TONGUE, etc. Descriptive terms for various features of lava tube caves.

LAVAFALL. A solidified lava cataract.

LAVA SEAL, LAVA SIPHON. A point where a lava tube is filled to the ceiling with lava, which hardened in place.

LAVA TRENCH. A long, narrow gulch that is a collapsed or never-roofed segment of a lava tube.

LAVA TUBE CAVE. A penetrable segment of an abandoned conduit of pahoehoe lava.

LAVA TUBE GLAZE. A shiny, relatively smooth coating of some lava tube caves.

LILY PAD. A special form of shelfstone formed around a stalagmite that has largely been submerged.

LIMESTONE. A sedimentary rock largely or completely formed of calcium carbonate.

LIMESTONE CAVE. A loose term for a solutional cave in almost any kind of rock.

LITTORAL. Pertaining to the zone between high- and low-water marks on a beach or cliff. "Littoral" caves are formed in this zone.

LIVE CAVE. A cave in which speleothem deposition is in progress.

MARBLE. Limestone that has been recrystallized and often molded by heat and pressure deep in the earth.

MASTER CAVE. A locally dominant throughway or trunk corridor.

MEANDER. (1) A segment of the course of a twisting stream; (2) a curved solutional channel incised into the wall or floor by such a stream.

METEOROLOGY. The study of the processes and contents of air and related phenomena.

MICROGOUR. A tiny rimstone dam on flowstone, on bedrock walls, or elsewhere in caves.

MIRABILITE. A hydrated sodium sulfate mineral.

MOGOTE. A term used especially in Cuba for conical hills typical of tropical karst.

MOON MILK. A white, puttylike form of flowstone, formed by one of several spelean minerals.

MOULIN. A domepit-like structure of glaciers.

NIFE CELL. A common type of wet-cell unit for electric headlamps.

OOLITE. A small rounded or faceted concretion.

OPALITE. A mineral composed of silicon dioxide, sometimes forming dripstone, flowstone, and rimstone.

OULOPHOLITE. A curved, fibrous gypsum crystal or group thereof.

PAHOEHOE. The relatively smooth-surfaced, once-fluid type of basaltic lava in which lava tubes form.

PALETTE. A broad, thin, disc-shaped speleothem. Dripstone often hangs from the margin.

PEPINO. A Puerto Rican term for *mogote*, especially elongated on amples.

PETROMORPH. A cavern feature exposed by solution of surrounding limestone (i.e., boxwork).

PHREATIC. Pertaining to the zone of water below the water table. In the phreatic zone, all cavern passages are filled with water, but the concept is poorly applicable to dense limestones.

PILLAR. A solitary vertical or near-vertical bedrock remnant. Cf. COLUMN.

PIPING. Progressive up-flow disintegration of a grainy rock along a water seep. Piping caves are formed by this process, sometimes with the assistance of solifluction.

PIRACY. "Capture" of a stream of water or lava by the drainage pattern of another.

PISOLITE. See OOLITE.

PIT. A natural shaft, large enough to descend.

PITCH. A specific length of vertical or near-vertical wall.

PITON. Ordinarily a thin, wedgelike blade that rock climbers hammer into cracks for attachment of carabiners and other climbing hardware. Snow and ice pitons and other atypical types also exist.

POCKETS, WALL OR CEILING. See JOINT POCKET (although these speleogens are not necessarily along joints).

PONOR. The point of disappearance of a stream in karstic topography.

POTHOLE. (1) A round, bowl-like pocket in the floor of a cave or stream; (2) a British term for pit or deep sink, occasionally used in Canada.

PRIMUS STOVE. A compact, lightweight mountaineers' stove.

PRUSIK KNOT. An ascent knot popularized by Dr. Karl Prusik, essentially a special use of the sliding hitch. Also useful for auto-belays and in other spelean situations discussed in various parts of the text.

PSEUDOKARST. Karstlike phenomena of glaciers, lava flows, and other poorly soluble rocks.

PURGATORY CAVE. A cave formed by accumulation of talus at the bottom of a narrow gorge.

RAFT. (1) A speleothem that floats on the surface of some cavern pools until it either adheres to the side or becomes too heavy for continued suspension by surface tension; (2) any shallow-draught conveyance that permits a caver to stay partially or entirely atop an underground pool.

RANDOM ELIMINATION. Defecation away from sanitary facilities.

RAPPEL. Controlled descent of a rope using friction of the body or a rappel device or both.

RAPPEL RACK. A U-shaped holder for several brake bars.

RECOVERY. Removal of a body after a fatal accident.

RESOLUTION. Solution of speleothems.

RESURGENCE. The point of surface appearance of a karstic stream.

RESUSCITATION. Restoration of heartbeat and breathing.

RIBBING. Vertical projections along the walls of domepits and between the channels of lapies.

RIBBON. An unbanded drapery that would otherwise be termed bacon-rind.

RIMSTONE. (1) Thin mineral crusts formed at the rims of some cavern pools; (2) terraced spelean deposits of calcite or other minerals, forming a complex of small or large basins.

ROCKFALL. The process of breakdown.

ROCKSHELTER. An overhung cavity that does not extend into total darkness. Often erroneously termed "cave."

SALTPETER. Speleologically, cavern deposits of nitrate minerals, usually in earthy deposits.

SALTPETER CAVE. A cave containing saltpeter, especially one formerly the site of saltpeter mining.

SCALLOP. An unevenly rounded shallow pocket of bedrock, glacier, or the like, occurring in groups. Those in limestone caves are sometimes called "stream flutes."

SEA CAVE. See LITTORAL.

SEAM. See WELT.

SEDIMENT. Material borne and deposited by water.

SEDIMENTARY ROCK. Rocks deposited in layers through the action of water; includes limestone and similar rocks of chemical origin, as well as those of lithified clastic sediments.

SELF-BELAY. See AUTO-BELAY.

SEWER PASSAGE. A comparatively small, roughly tubular passage, which intermittently or continuously transmits large quantities of water.

SHAKEHOLE. A British term for "sinkhole." Sometimes encountered in Canada.

SHELFSTONE. Extensive mineral shelves formed at the rims of some cavern pools. See also RIMSTONE.

SHIELD. See PALETTE.

SILICEOUS. Pertaining to or formed of silica minerals.

SINK. A rounded depression without a surface outlet.

SINKHOLE. Essentially identical with sink.

SIPHON. Obstruction of a section of cavern by water or lava that reaches to the ceiling.

SLACK. Lack of tautness in a rope.

SLING. A vertical caver's short length of rope or webbing.

SNOW CAVE. A natural cave in snow formed by the same processes which create glacier caves. See also FIRN CAVE. (*Note:* mountaineers' "snow caves" are artificial bivouac holes.)

SODA-STRAW STALACTITE. A thin-walled, hollow, tubular stalactite, approximately the diameter of a drop of water.

SOLUTIONAL TUBE. A round or oval, comparatively straight passage, often too small for human entry.

SOTANO. A Spanish-language term for "pit," used in much of Spanish America. Also applied to some deep sinks.

SPAN. A spelean natural bridge.

SPAR, DOGTOOTH, NAILHEAD, RICE CRYSTAL, etc. Types of small calcite crystals, vaguely resembling their namesakes.

SPELEAN. Pertaining to caves.

SPELEOBIOLOGY. Cave biology.

SPELEOGEN. A cave feature resulting from removal of bedrock.

SPELEOGENESIS. The process of origin and development of caves. The corresponding adjective is "speleogenetic."

SPELEOLIFEROUS. Containing caves (applied to certain limestones, lava flows, etc.).

SPELEOLOGY. The study of caves and related phenomena.

SPELEOTHEM. Any mineral deposit formed in a cave.

SPELUNKER. A caver.

SPELUNKING. Sport caving.

SPONGEWORK. See HONEYCOMBING.

SQUEEZEWAY. A cavern passage so narrow that human progress is difficult.

STACKED PASSAGES. A succession of vertically aligned passages.

STANDING ROPE. A rope that is securely anchored and allowed to dangle.

STATIC BELAY. A type of belay that seeks to halt a fall immediately, without allowing the rope to run.

STEAM CAVE. See GEOTHERMAL CAVE.

STERNO. A commercial flammable chemical sold in cans (tins).

STOKES STRETCHER. A common rigid rescue stretcher.

STREAM FLUTE. See SCALLOP.

SWALLET or SWALLOW HOLE. The British equivalent of *ponor*, moderately used in Canada.

SWEEP. A comparatively rapid, systematic cave search.

SWISS SEAT. A body harness loop of webbing encompassing the waist and gluteal regions, held together in front by a carabiner.

SYNCLINE. A trough-shaped or down-arched bedrock fold. See also ANTICLINE.

TALUS. Large and/or small rocks displaced from their original position.

TECTONIC. Pertaining or related to internal movements of the earth.

THERMAL. See GEOTHERMAL.

THROUGHWAY. A large, near-horizontal, comparatively straight passage with uniform characteristics for hundreds or thousands of feet.

TRANSITION ZONE. The part of a cave where temperature changes are perceptible but significantly less than in the entrance zone. Often this is approximately the same as the twilight zone.

TRAVERTINE. Speleologically, a coarse form of flowstone or rimstone, often of organic origin. Sometimes applied to any calcium carbonate speleothem.

TREE CAST. A mold of a tree engulfed in lava, usually pahoehoe but occasionally in pillow lava or other volcanic rocks.

TROGLOBITE. An animal that lives exclusively in caves.

TROGLODYTE. A nonspecific term for cave dweller, human or otherwise.

TROGLOPHILE. An animal that regularly lives in total darkness of caves but can and does survive outside if conditions are favorable.

TROGLOXENE. An animal that occasionally visits or temporarily lives in caves.

TRUNK PASSAGE or CHANNEL. See THROUGHWAY and MASTER CAVE.

TUBE-IN-TUBE. A rudimentary lava tube formed in a secondary flow inside a larger lava tube.

TYROLEAN TRAVERSE. A dramatic mountaineers' method of getting across a chasm by lassoing a pinnacle on the far side, then tightening the rope and pulling oneself along its length.

VADOSE. Pertaining to subsurface water in the zone above the water table.

VANDALISM. Ignorant or deliberate damage to cave resources and values. See also CONSERVATION.

VUG. A crystal-lined underground cavity, normally too small to be termed a cave.

VULCANOSPELEOLOGY. The study of lava tube caves and related phenomena.

WATER CHILL. Impending or actual hypothermia as a result of exposure to water.

WATER TABLE. The upper surface of the zone saturated with ground water. The concept is poorly applicable to massive limestone.

WELT. A seamlike speleothemic deposit along the course of a joint or a crack in a speleothem.

WET SUIT. A diver's suit penetrated by water yet able to maintain a thin layer of warmth against the body because of the insulation of innumerable bubbles of water trapped in its construction.

WHALETAIL. A solid rappel device with slots that permit incremental control.

WICKING. The process by which capillary action of wet clothing draws body heat to the surface, where it is lost.

WIND CHILL. Impending or actual hypothermia as a result of exposure to wind currents.

Suggested Additional Reading

CHAPTERS 1, 2, AND 3:

Fundamental to scientific cave exploration in North America is J Harlen Bretz's "Vadose and Phreatic Features of Limestone Caverns" (*Journal of Geology*, vol. 50, no. 6, pt. 2, 1942). His *Caves of Missouri* (Missouri Geological Survey, 1956) and other works provide additional data. William M. Davis published an even more fundamental deductive article, "Origin of Limestone Caverns," in the *Bulletin* of the Geological Society of America (vol. 41, no. 3). Innumerable subsequent articles have appeared in speleological and other scientific literature. *Current Titles in Speleology* (published yearly in Great Britain) and the National Speleological Society's *Speleodigest* serve as guides. Joe Jennings's *Karst* (Australia National University Press, Canberra, 1971) is a fine introduction to the subject. Marjorie Sweeting's world-renowned *Karst Landforms* (Columbia University Press, New York, 1973) has a British slant.

The classification of cave features in this book is my own (NSS *Bulletin*, vol. 17, 1955). That of speleothems of limestone caves is modified from that in my *Caves of California* (Western Speleological Survey, Seattle, 1962). Carol Hill's *Cave Minerals* (NSS, 1976) is a thorough coverage for American caves. Clastic fills and rockfall are particularly well considered in the works of William E. Davies, including *Caves of West Virginia* (West Virginia Geological Survey, 3 editions), *Caves of Maryland* (Maryland Department of Geology, 2 editions), and articles in the NSS *Bulletin* (especially in vol. 11, 1949, and vol. 13, 1951). A more recent key article by White and White is in vol. 31, no. 4, of the same publication.

Data on specific limestone caves mentioned in these chapters is scattered far and wide in such publications as the Huntsville Grotto *Newsletter*. The expanded 1976 edition of my *Depths of the Earth* (Harper & Row, New York) tells the stories of many of them. The

Johnson Reprint Company's 1970 republication of the Reverend Horace C. Hovey's 1896 *Celebrated American Caverns* provides perspective on the progress of American caving and speleology. *The Longest Cave* by Roger Brucker and Richard A. (Red) Watson (Knopf, New York, 1976) describes the exploration of the Flint Ridge system and its eventual connection to Mammoth Cave. The 1981 8th International Congress of Speleology *Guide-Book to the Historic Section of Mammoth Cave* is the latest word on that subject. The *Manual of Caving Techniques* published by the Cave Research Group of Great Britain in 1969 (Routledge and Kegan Paul, London) provides excellent data on cave digging. Sheck Exley's *Basic Cave Diving: a blueprint for survival* (NSS, two editions to date) should be studied by all would-be cave divers. *Underwater Speleology* is the newsletter of the section on cave diving of the NSS. Jack Burch discussed advanced cave-finding techniques in the August 1968 issue of *Down Under,* newsletter of the National Caves Association. Pseudokarst is discussed in my 1960 article in the NSS *Bulletin* (vol. 22, pt. 3) and followup articles. Vulcanospeleology has been the subject of increasing study. The *Proceedings* of the 1972 International Symposium on Vulcanospeleology and Its Extraterrestrial Applications (Western Speleological Survey, Seattle) provide a considerable overview.

Several important fissure, talus, and block-creep caverns in the eastern United States are described in Clay Perry's *Underground Empire* and *New England's Buried Treasure,* both published by the Stephen Daye Press in the 1940s. Important examples in Idaho are included in Sylvia Ross's *Introduction to Idaho Caves and Caving* (Idaho Bureau of Mines and Geology, Moscow, 1969).

Additional information on hypoxia and "bad air" can be found in Lundy's *Clinical Anesthesia* (W. B. Saunders, Philadelphia, 1942), Rosenau's *Preventive Medicine* (Appleton-Century, New York, 1965), Armstrong's *Principles and Practice of Aerospace Medicine* (Williams and Wilkins, Baltimore, 1971), Best and Taylor's *Physiological Basis of Medical Practice* (Holt, Rinehart and Winston, New York, 1966), Latimer and Hildebrand's *Reference Book of Inorganic Chemistry* (Macmillan, 1964), and similar multiedition texts. In 1961 *The Texas Caver* printed a two-part article on "bad air caves" that summarizes the problem.

Basic data on cave hydrology and meteorology are summarized in Moore and Nicholas's *Speleology: The Study of Caves* (Zephyrus Press, Teaneck, N.J., 1978), albeit without due consideration of alternate concepts in some important matters. My 1954 "Ice Caves of the United States" (NSS *Bulletin,* vol. 16) was brought up to date in the 28-page introduction to the 1970 Johnson Reprint Company edition of E. S. Balch's classic 1900 *Glacières, or Freezing Caverns.*

Herb Conn prepared a fascinating 1966 article on barometric wind in "Wind and Jewel Caves, South Dakota" (NSS *Bulletin,* vol. 28, no. 2). He and Jan also somehow found time to write a delightful little book: *The Jewel Cave Adventure* (Zephyrus Press, Teaneck, N.J., 1976). My *Depths of the Earth* is a good starting point for several other meteorological phenomena. A progress report on the Burnsville cave system appeared in the July 1971 NSS *News.*

The Sierra Club has epitomized conservation education in such magnificent books as *The Wild Cascades* (1965) and *Time and the River Flowing* (1964). Those skeptical of the power of the informed American conscience might well ponder the fate of Glen Canyon, subject of its posthumous *The Place That No One Knew* (1963, 1966). A generation ago, I knew and loved Glen Canyon. Standing alone, six of us came heartbreakingly close to overturning the then-undefeated Bureau of Reclamation political steamroller that destroyed its unique resources. Quite possibly we were not much more than one such book from success. With books and other educational media, the swelling ranks of conservationists soon mobilized public opinion to save the Grand Canyon, enact the Wilderness Bill, create the North Cascades National Park, and much more. The closest American book to this tradition is Lloyd E. Parris's 1973 *Caves of Colorado* (Pruett Publishing Co., Boulder, Colo.). Would that several Sierra Club books looked beneath the surface!

CHAPTERS 4 AND 5:

The *Personnel Manual* of the Cave Research Foundation is a particularly valuable source for many of the topics covered in these chapters.

More succinct is Jack Stellmack's analysis of a week-long underground camp-out published in the February 1968 *Nittany Grotto News*. The Mountaineers' noted *Freedom of the Hills* (Seattle, 3 editions to date) is another outstanding reference source.

Information on carbide lamps is scattered widely through the speleological literature. Donald Davis described techniques of repairing old headlamps in the August 1970 NSS *News*.

Further information on electric caving can be found in various reports by Bill Varnedoe, especially in the October 1970 *Bulletin* of the National Speleological Society and the *Huntsville Grotto Newsletter*, especially in its March 1965 and August–October 1967 issues, which include key data on candlepower and other parameters. Many others have also contributed. Especially useful background information was published in the January 1971 *Popular Science* and July 1970 *Field and Stream*. Information on British-made wet-cell units is well summarized in the Cave Research Group's *Manual of Caving Techniques*. Catalogs of such manufacturers as the Chicago Miniature Lamp Works contain useful background information. Mountain Safety Research *Newsletter & Catalog* no. 11 (March 1976) describes the characteristics of Lithium D-cells in more detail than recent issues. Advanced helmet-mounted and other electric systems are discussed in *Northwest Caving* (vol. 10, no. 1, Summer 1980) and *Proceedings* of the 1981 NSS Northwest Regional Symposium on Cave Science and Technology, including considerations of 6 volt, 1 amp/hour quartz-halogen bulbs.

Fundamental data on hypothermia were published as a 168-page U.S. Navy book in 1949. By H. B. Eisberg, it is entitled *Arctic and Cold Weather Medicine*. The Mazamas of Portland, Oregon, have published an excellent 13-page booklet entitled *Hypothermia: Killer of the Unprepared*. The Mountain Rescue Association, too, has issued several excellent publications on the subject and a fine 16-mm. movie, *By Nature's Rules*. It is now widely available. Marlin Kreider is one of the few on this continent who have applied these principles specifically to caving (National Speleological Society *Bulletin*, vol. 29, no. 1, 1967). A three-part article by Dr. Warren C. Hunter in the June, August, and September 1968 issues of *Northwest Medicine* contains other important information. An excellent recent summary appeared in *Contemporary Orthopedics* (vol. 2, no. 3, June 1980). A similar summary of exertional heat injury was published in *The Medical Letter* in July 1981 (vol. 23, no. 14). The *Proceedings* of the 1981 Northwest Regional Association Symposium on Cave Science and Technology analyzes several current makes of helmets, ropes, harnesses, ascenders, etc. (see below). The Philadelphia Grotto of the NSS has published especially good items on cave food in its *Digest* (vol. 4, no. 2 and vol. 7, nos. 1–2).

CHAPTERS 6, 7, AND 8:

Many articles on various vertical caving techniques are scattered through the recent North American caving literature. Although in confusing "British English," Neil Montgomery's *Single Rope Techniques* (Sydney Speleological Society, Sydney, Australia, 1977) is a good overview. Bob Thrun's *Prusiking* (NSS, 2 editions to date) also is a basic text, although a bit outdated. *Nylon Highway* (newsletter of the Vertical Section of the NSS) and *Caving International* contain especially current articles. Fred Wefer's article and notable graphs appeared in the March–April 1970 *Nittany Grotto News*. Carl Kunath published a short analysis on determining the depth of pits by audio methods in the *Texas Caver* (vol. 10, no. 6). The cover of the January 1981 *Cleve-o-Grotto News* featured a "ropewalker" climbing harness; other diagrams appear in the catalog of Gibbs Products (see list of equipment and information sources). Failures of the British-made Lewis ascender (plus unique problems with a rappel rack) hundreds of feet above the floors of deep Mexican sotanos are mentioned in Jim Eyre's dramatic *The Cave Explorers* (Stalactite Press, Calgary, Alberta, 1981). *Freedom of the Hills* contains valuable information on the use of pitons, other hardware, and basic climbing techniques. Royal Robbins's 1971 *Basic Rockcraft* and his 1973 *Advanced Rockcraft* (both published by La Siesta Press, Glendale, Calif.) are valuable booklets. The Sierra Club's famous 1946 *Belaying the Leader* remains a classic. Those who read French will enjoy *Techniques de la Spéléologie Alpine* (two editions published in France by Techniques Sportives Appliquées) despite major differences between American and European equipment. Most public libraries have several books on knots.

Cavers, however, may want to order Wheelock's *Ropes, Knots, and Slings for Climbers* from mountaineering stores. *Newsletter* no. 3 of Mountain Safety Research (1970) discussed the friction hitch and Stitcht belay plate, including use of the latter in rappelling. Its first issue included the MSR auto-belay gadget. Cave telephones are especially well discussed in the British book *Manual of Caving Techniques* (see above). Dave Mischke's simple telephone system was illustrated in the May 1971 NSS *News*. Use of radios in and around caves was the subject of Western Speleological Survey *Serial* no. 64 (1981).

CHAPTERS 9 AND 10:

Basic publications on first-aid techniques are available from the American Red Cross and other organizations. Many mountaineering books contain modifications of these techniques that may be useful to cavers. *Mountaineering First Aid,* by Dick Mitchell of Huntsville and Seattle, and Dr. Fred Darvill's pocket-sized *Mountaineering Medicine* are recommended especially. Further data on midair resuscitation can be obtained in the August 1971 *Northwest Medicine* (vol. 13, no. 8). Although now somewhat dated, my *Cave Sicknesses* (NSS *Bulletin* vol. 11) and *Medical Hazards to Cave Explorers* (GP, vol. 18, no. 1, 1958) provide basic guidelines. Current information on histoplasmosis and coccidioidomycosis can be found in the 1980 *Handbook on Fungus Diseases* published by the American College of Chest Physicians; a short overview of the latter is in *Basics of Respiratory Disease* (vol. 9, no. 2, November 1980). Improved rabies immunization is discussed in the August 8, 1980, issue of the *Journal* of the American Medical Association and elsewhere; it appears that it still causes some serious reactions. Valuable data substantiating the need to protect, not destroy, cave bats appears in an article on skunk-associated human rabies in the October 2, 1972, issue of the *Journal* of the American Medical Association and in the NSS *Bulletin,* vol. 34, no. 2 (the proceedings of a symposium on the ecology, physiology, behavior, and survival of bats). Wind chill tables are published frequently in *U.S. News and World Report* and REI catalogs. Medical problems resulting from caving and other exertion at high altitudes are summarized in the January 23, 1981, issue of *The Medical Letter*. A good recent article on carbon monoxide poisoning appeared in the October 9, 1981, issue of the *Journal* of the American Medical Association. The treatment of rattlesnake bite continues to be controversial; Russell's *Snake Venom Poisoning* (Lippincott, Philadelphia, 1980) is widely accepted. Regarding tropical cave diseases, discussions in the April 18, 1980, *Journal* of the American Medical Association provide a beginning on a topic which needs further study. A good overview of schistosomiasis appeared in the July 1980 *Western Journal of Medicine* (vol. 133, no. 1). Water purification was discussed in the same journal in August and October, 1981. Information on the bite of the brown recluse spider can be found in the April 1972 *Northwest Medicine* (vol. 71, no. 4).

The report of the American Medical Association's Committee on Exercise and Physical Fitness appeared in that association's *Journal* on February 14, 1972. The American Heart Association is publishing a series of handbooks on exercise training and testing. Additional recent medical data on conditioning can be found in the August 1, 1980, July 17, 1981, and August 28, 1981, issues of the *Journal* of the American Medical Association, the June 1980 issue of *Contemporary Orthopedics*, and in recent books by Lawrence Cohen, Paul Fardy, Michael Pollock, and others.

Regarding supposedly erogogenic drugs, a recent overview appeared in the December 1980 *Western Journal of Medicine* (vol. 133, no. 6). A useful similar overview of marijuana was in the October 16, 1981, *Journal* of the American Medical Association. Street drug misrepresentation was the subject of an article in its July 25, 1980, issue.

Jogging techniques are outlined in publications of the National Jogging Association (P.O. Box 19367, Washington, D.C. 20036), including Dr. Richard Duhannon's 1970 *Guidelines for Successful Jogging*, Bill Cuddington especially recommends *Runners World* (P.O. Box 366, Mountain View, California 94040) and the Coopers' recent *The New Aerobics* and *Aerobics for Women* (Bantam Books and Lippincott, respectively).

William B. (Skeets) Miller's first-person account of the Floyd Collins fiasco appeared in the *Reader's Digest* in April 1962. Many of the published accounts of the episodes are unreliable, and some are so incorrect that the Collins family has successfully sued the writers and publishers for libel. My chapter thereon in *Depths of the Earth* is based on especially intensive research. *Trapped,* by Roger Brucker and Robert K. Murray (Putnam's, New York, 1979) is an interesting analysis of the rescue attempt.

The 1981 *Manual of U.S. Cave Rescue Techniques* (NSS National Cave Rescue Commission) contains much valuable information. Despite some odd misconceptions, the spelean chapter of the Appalachian Mountain Club's *Wilderness Search and Rescue* (Boston, 1980) has many practical suggestions also. Additional information can be obtained in *Freedom of the Hills,* and in the Cave Research Group's *Manual of Caving Techniques.* The NSS also publishes periodic reports entitled *American Caving Accidents.* Terry Tarkington's delightful account of the Anvil Cave caper appeared in the December 1967 *Huntsville Grotto Newsletter.* Organization of the Hannibal search was described in particular detail in the December 1967 issue of the *Speleologist*; the Twiggs Cave rescue in the March 1978 NSS *News.* Perhaps the best current information on new safety and rescue equipment is the "New Products and Ideas" section of *Occupational Health and Safety,* published monthly. The impact of caves on non-caver rescue teams is discussed in Tom Vines's "Cave Rescue Management" (*Emergency,* June 1981).

GENERAL REFERENCES

For other information on caves of the United States and some other parts of North America, the reader may wish to refer to the similar section of my *Depths of the Earth* (Harper & Row, New York, 1976). The following also may be of interest:

Barr, T. C. 1961. *Caves of Tennessee.* Tennessee Division of Geology, Nashville.

Beck, Barry F. 1980. *An Introduction to Caves and Cave Exploring in Georgia.* Georgia Department of Natural Resources, Atlanta.

Bretz, J Harlen, and Harris, S. E., Jr. *Caves of Illinois.* Two editions. Illinois Dept. of Registration and Education, Urbana.

Campbell, Newell P. 1978. *Caves of Montana.* Montana Bureau of Mines and Geology, Butte.

Dalton, Richard F. 1976. *Caves of New Jersey.* New Jersey Dept. of Environmental Protection, Trenton.

Douglas, H. H. 1964. *Caves of Virginia.* Virginia Cave Survey, Falls Church.

Franz, Richard, and Slifer, Dennis. 1971. *Caves of Maryland.* Maryland Geological Survey, Baltimore.

Hauer, Peter M. 1969. *Caves of Massachusetts.* Speece Publications, Altoona, Penna.

Hill, Chris; Sutherland, Wayne; and Tierney, Wayne. 1976. *Caves of Wyoming.* Geological Survey of Wyoming, Laramie.

Hogbert, R. K. and Bayer, T. N. 1967. *Guide to the Caves of Minnesota.* Minnesota Geological Survey, Minneapolis. (Mostly on commercial caves.)

Larson, Charles V. 3rd edition in press. *Caves of Oregon.* Western Speleological Survey, Vancouver, Wash.

Matthews, Larry E. 1971. *Descriptions of Tennessee Caves.* Tennessee Division of Geology, Nashville.

Northeastern Regional Organization of the National Speleological Society. 1979. *New Hampshire Caves* (preliminary edition).

Powell, Richard L. 1961. *Caves of Indiana.* Indiana Geological Survey, Bloomington.

Scott, John. 1959. *Caves in Vermont.* Killooleet, Hancock, Vt.

Speece, Jack. 1977. *The Caves of Delaware.* Speece Productions, Altoona, Penna.

Yale Grotto of the National Speleological Society. 3 editions to date. *Caves of Connecticut.*

Beginning in 1960 with Carlsbad Caverns National Park, National Speleological Society convention guidebooks are especially helpful. Statewide issues of its *Bulletin* include Texas (*Bulletin* 10), Pennsylvania (*Bulletin* 15), Michigan (an article by William E. Davies in *Bulletin* 17) and Hawaii (my cursory article in *Bulletin* 20).

Canadian caves and caving are reported periodically in *The Canadian Caver* (see list of equipment and information sources). This magazine also published a notable book in 1976: *Cave Exploration in Canada.* Several other Canadian speleological organizations publish worthwhile newsletters.

Mexican caving is considered in publications of the Association for Mexican Cave Studies (see list of equipment and information sources) and many other American and Canadian speleological publications. Two Mexican groups have published newsletters or bulletins, but these are not widely available north of the Rio Grande.

Caves of Guatemala and Belize have been discussed in several recent issues of the NSS *Bulletin,* the *Explorers Journal, Caving International,* and other publications. Those of Cuba have been described in several systematic recent works of Antonio Nuñez Jiminez and others, not widely available in this country. Many older books and articles about the island contain information on Bellamar and other caves. Thomas Barbour's *A Naturalist in Cuba* (Little, Brown & Co., Boston, 1945, 317 pp.) is particularly pleasant. Justice has not yet been done to the caves of Puerto Rico and many other portions of the West Indies. NSS *Bulletin* vols. 29, no. 2, and 30, no. 3, contain a beginning for Puerto Rico. Fincham's 1977 *Jamaica Underground* (Geological Society of Jamaica, Kingston) is a fine resource for that island.

And I especially recommend Robert Nymeyer's *Carlsbad, Caves, and a Camera* (Zephyrus Press, Teaneck, N.J., 1978).

Equipment and Information Sources

Association for Mexican Cave Studies. P.O. Box 7672, UT Station, Austin, Texas, 78712. Publications and information on Mexico and Central America.

Bluewater Ltd. 131 Lavvorn Road, Carrollton, Ga. 30117. The famous caving ropes and other selected vertical gear. Free price lists.

Bob and Bob Enterprises. P.O. Box 441, Lewisburg, W. Va. 24901. The best-known American speleo-supplier. Free price lists.

Bright Star Industries. 600 Getty Avenue, Clifton, N.J. 07015. Regional warehouses in Chicago, Dallas, Atlanta, San Francisco, and Los Angeles. Industrial flashlights and batteries. Free information.

Canadian Caver. P.O. Box 275, McMaster University, Hamilton, Ontario, Canada L8S 1C0. Publication with scope far beyond Canada.

Cave Research Foundation. Frequent address changes; query the NSS (see below).

Caving International. P.O. Bag 4014, Station C, Calgary, Alberta, Canada T2T 5M9. Outstanding publication with information on equipment and techniques as well as feature articles.

Donald G. Davis. P.O. Box 25, Fairplay, Colorado, 80440. Premier headlamps and parts. Free price list and information sheet.

Early Winters, Ltd. 110 Prefontaine Place South, Seattle, Wash. 98104. Lithium cells and lighting systems of interest to cavers. Free catalog.

Gibbs Products. 202 East Hampton Avenue, Salt Lake City, Utah, 84111. Gibbs ascenders plus select vertical gear. Free price list and information.

Grotto Store. 12 South Main Street, Rittman, Ohio. 44270. General stock of caving equipment. Free price list.

Herbach and Rademan, Inc. 401 East Erie Avenue, Philadelphia, Penna. 19314. Gates, G.E., and other lighting systems. Free catalog.

Inward Bound. 378 Webster Street, Manchester, N.H. 03104. Wet and dry suits and accessories. Free catalog.

Mountain Safety Research. 631 South 96th Street, Seattle, Wash. 98108. Lithium cells, special headlamps (on headbands) and other special gear of interest to cavers. Free catalog.

National Cave Rescue Commission. Mailing address: c/o NSS (see below). 24-hour telephone number: 1–800–851–3051.

National Speleological Society. Cave Avenue, Huntsville, Alabama, 35810. Telephone (work hours): 1–205–852–1300. Free information.

Pathfinder Sports, Inc. 5214 East Pima Street, Tucson, Ariz. 85712. Especially broad line of rock-climbing gear and some caving gear including chocks and hard-to-find light systems. Free catalogs.

Pigeon Mountain Industries. P.O. Box 803, Lafayette, Georgia, 30728. The popular PMI rope, plus rope washers and other vertical gear. Free price list and information.

Recreational Equipment, Inc. (REI) P.O. Box C-88125, Seattle, Wash., 98188. Stores in Seattle, Portland, Los Angeles, Berkeley, Bloomington (Minn.), and Anchorage. Broad line of rock-climbing gear. Free catalogs.

Safesport Manufacturing Company. 1100 West 45th Avenue, Denver, Colo. 80211. Safesport carbide headlamps and parts. Free price list.

Sierra Club. Mills Tower, San Francisco, Calif. 94104. Conservation and wilderness travel information. Free information.

Sociedad Espeleologica de Puerto Rico. Apartado Postal 31074, 65th Inf. Sta., Rio Piedras, P.R. 00929. Liaison with Puerto Rico cavers.

Sociedad Venezolana de Espeleología. Apartado 47,334, Caracas, Venezuela 1041-A. Publications and information on caving in much of Spanish America and the Caribbean.

Speleobooks. P.O. Box 333, Wilbraham, Mass. 10195. New and out-of-print books and ephemera.

Speleoshoppe. P.O. Box 297-N1, Fairdale, Ky. 40118. Large stock of caving gear including some European items.

Vertical Cave Equipment and Supply. 23 Charcoal Ridge Road South, Danbury, Conn. 06810. Small line of key rock-climbing and vertical caving gear. Free catalog.

Western Speleological Foundation. 1117 36th Avenue East, Seattle, Wash. 98112. Nonprofit organization advancing the systematic study, interpretation, and preservation of caves and related natural and cultural phenomena. Free information.

Western Speleological Survey. 13402 N.E. Clark Road, Vancouver, Wash. 98665. Technical publications.

Wilderness Society. 729 15th Street N.W., Washington, D.C. 20005. Conservation publications. Free information.

Acknowledgments

So many people helped me prepare *Adventure Is Underground* and *Depths of the Earth* that it was impossible to thank them all properly. This time, the situation is even worse. After compiling this book, I am more than ever aware of my obligation to the thousands of cavers and speleologists who have made our sporting science—or our scientific sport, if you prefer—what it is today. People like Bill Cuddington and Bill Stephenson—to whom I am honored to be able to dedicate this work—not only for their role in developing modern cave exploration in North America, and for their specific assistance in providing information unobtainable elsewhere, but also for their unflagging friendliness and effective teaching.

People like J. S. Petrie, who kept the National Speleological Society alive when all seemed lost during World War II.

And so many others, living and dead. Bill Cuddington, Ome Daiber, Carol Hill, Dave Mischke, Bill Varnedoe, and Will White for generously and patiently contributing their special experience by critically reviewing early drafts of sections of the book—and much else. Charley Anderson, for his inspiration and dedication to glaciospeleology and—with Mark Vining—for most of the drawings in this book. J Harlen Bretz for teaching us how to read the language of caves. John Bridge and many others in the Cave Research Foundation inside and outside Flint, Joppa, and Mammoth Cave ridges. Bob Brown, for his leadership in the Cascade Grotto and the Northwest Cave Rescue Association. The entire Cascade Grotto and the Northwest Region of the National Speleological Society, for untold pioneering in vulcanospeleology and glaciospeleology over the years. Denny Constantine, for long-continued helpfulness in bat rabies and other matters. Chuck Coughlin, for information on the Summit Steam Caves of Mount Rainier and much besides. Don Cournoyer, for reports on more than one hundred trips into Virginia's Breathing Cave. The Craven Pothole Club, and especially Don Mellor, for a splendid taste of caving outside North America.

Ome Daiber, for being "Mr. Mountain Rescue" and for being so willing to share his vast experience so widely. Bill Davies, for his notable contributions to speleology and for putting up with my kidding in a previous book. Roy Davis, for many a yarn and many a precise fact about Cumberland Caverns. Burton Faust for his historical research and the rare gift of contagious dedication. Bill Foster for his initial impulse to cave mineralogy in the United States. Jerry Frahm, for eternal helpfulness. Verne Frese, for patient, unsung effort in the interest of Northwest cave rescue. Dr. Michael Furcolow, for his patience in putting up with the correspondence of a brash young physician-in-training who was concerned about histoplasmosis.

Frank Hjort, for "inside information" on lava tubing by the staff of Hawaii National Park while the lava was still warm. Frank Howarth, for similar inside information on cooled-off but undescribed Hawaiian lava tube caves. Clarence Hronek, for developing Vancouver Island caving and for information on various other Canadian caves. George F. Jackson, about whom I could write a book, and may sometime.

Still others, like Tom Meador, self-effacing and ever anxious to help. Dave Mischke, for advancing the rescue skills of all associated with him, and for developing an excellent cave telephone system, as well as for reviewing part of this book. Charles Mohr, for his long leadership in speleology and in cave conservation. Joaquin Montoriol, for clarifying many obscure principles of cave meteorology and clastic fills, and for assisting field work far from his Barcelona home. George W. Moore, for contributions to speleology, for assistance in the field, and for introducing the term *speleothem*. Rudy Mosco, for his account of the Sulfur Cave at Steamboat Springs. The Mountaineers, for their outstanding conservation efforts and their publication of such masterpieces as *Freedom of the Hills*.

Bobbi Nagy, for enlivening caving. The National Park Service, for service far beyond the call of duty. Ike Nicholson, of Butler Cave. Tony Oldham, for warm hospitality and assistance virtually the length of Britain. Garald Parker, for valuable exchange of ideas on pseudokarst. Ed Post, for exceptional assistance in historical research. Jim Schermerhorn, for leadership inside Arkansas. Vic Schmidt, probably America's number one cave conservationist. Ken Sinkiewicz and others in the Vancouver Island Cave Exploration Group, for warm assistance and overwhelming friendliness. Trevor Shaw, for so much that I wouldn't know where to begin. Carroll Slemaker, for a valuable first-person account of his rescue. Gordon Smith of Joppa Ridge. Roger Smith, Britain's counterpart to Vic Schmidt, and Anthony Sutcliffe, prime mover toward the Pengelly Cave Research Centre and a window to caving in half the world.

Bill Varnedoe, for being himself, besides reviewing a section of this book. The Virginia Polytechnic Institute Grotto, for hosting a magnificent mudhole that was still a capital convention of the National Speleological Society—and for compiling the trouble-shooting carbide lamp chart I included here in modified form. Jerry Vineyard, for years of magnificent contribution to Missouri speleology and every caver who came his way. Will White, for great contributions to speleogenetic theory, as well as reviewing part of this book. Phil Whitfield, for information on British wet-cell lamps not easily obtained otherwise.

And especially those who have contributed photographs to the book.

And my family and Harper & Row, neither of which is as vexed as they have every right to be after all these months of neglect.

To all these and the thousands more, my heartfelt thanks.

Illustration Credits

Drawings

Figs. 1–14, 19, 22–25:   William R. Halliday.
Figs. 15, 18:   Sketches by Bruce Smith, courtesy of Bill Cuddington.
Figs. 16–17:   Sketches by Bruce Scott, courtesy of Bill Cuddington.
Fig. 20:   Courtesy of Bill Varnedoe; redrawn by Charles H. Anderson, Jr., and Mark Vining.
Fig. 21:   Data derived from surveys by Max Kaemper, the Cave Research Foundation, and others. The plans of some passages have been simplified for clarity and others—especially those lying along the edge of ridges—necessarily omitted.

Photographs

Photograph credits are listed by page numbers. All other photographs courtesy of William R. Halliday.

143 William Brown.
151 Vic Schmidt.
154–55 Jack E. Boucher, courtesy National Park Service.
158 Basil Hritsco.
167 Luray Caverns, Virginia.
182 Dwight Deal.
201 Luray Caverns, Virginia.
210 Ed Yarbrough.
254 John H. Tuohy, M.D.
278 Carroll Slemaker.

Index